TELECOMMUNICATIONS

TELECOMMUNICATIONS

ELECTRONIC TELEVISION

By

GEORGE H. ECKHARDT

ARNO PRESS
A New York Times Company
New York • 1974

Reprint Edition 1974 by Arno Press Inc.

Reprinted from a copy in
The Newark Public Library

TELECOMMUNICATIONS
ISBN for complete set: 0-405-06030-0
See last pages of this volume for titles.

Manufactured in the United States of America

Library of Congress Cataloging in Publication Data

Eckhardt, George H
Electronic television.

(Telecommunications)
Reprint of the 1936 ed. published by Goodheart-Willcox Co., Chicago.
1. Television. I. Title. II. Series: Telecommunications (New York, 1974-)
TK6630.E4 1974 621.388 74-4674
ISBN 0-405-06040-8

ELECTRONIC TELEVISION

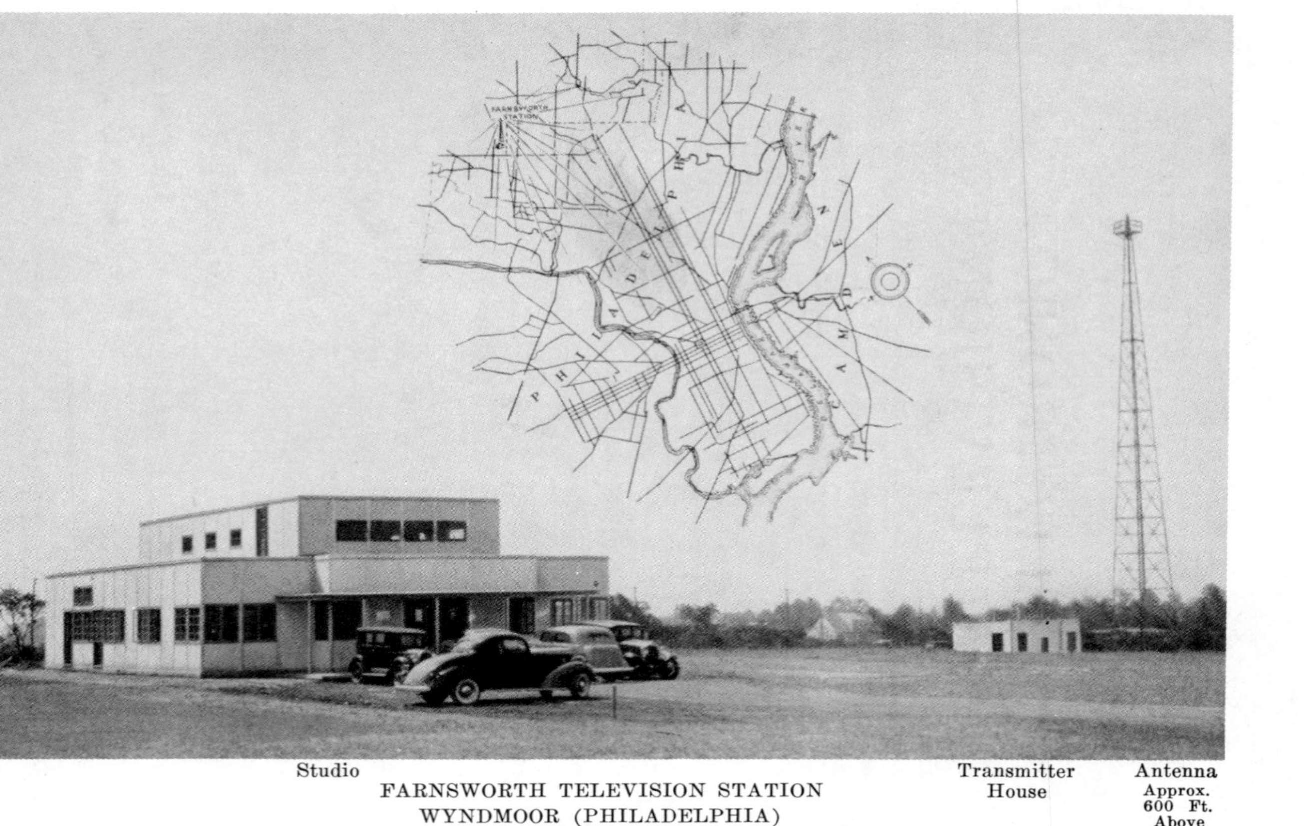

Studio   Transmitter House   Antenna Approx. 600 Ft. Above Sea Level

FARNSWORTH TELEVISION STATION
WYNDMOOR (PHILADELPHIA)
Diagram Indicates Approximate Directional Coverage of Station

ELECTRONIC TELEVISION

By

GEORGE H. ECKHARDT

CHICAGO

THE GOODHEART-WILLCOX COMPANY, INC.

Publishers

1936

Printed in the United States of America

PREFACE

Two men stand out as pioneers in electronic television—Philo T. Farnsworth, of Farnsworth Television Incorporated; and Dr. V. K. Zworykin, of the Radio Corporation of America. Practically all of the basic research in this work has been done in the laboratories of these two companies, and it would be impossible to conceive any new development that would not encroach upon the basic research done in one or the other of these laboratories. From time to time both Mr. Farnsworth and Dr. Zworykin, and members of their respective staffs, have contributed articles to proceedings of engineering and other learned societies, especially the *Journal of The Franklin Institute.* These articles have been scattered and were written primarily for advanced engineers and scientists.

With the emergence of electronic television from the laboratory into the field, it was obvious that an authentic book on the subject was needed, a book that not only would be accurate, but also would be written in a manner understandable to readers other than highly trained engineers.

It was equally obvious that the most reliable source of information would be the two laboratories in which the research had been done. The writer, therefore, wishes to express his profound appreciation to the Radio Corporation of America and Farnsworth Television Incorporated, for their splendid cooperation and patience. Both of these companies kindly consented to check over material in the book regarding their respective systems. Without their help, in matters pertaining to each of their respective systems, the book could not have been written, and it is doubtful whether any authentic book on electronic television could be written without their help.

The writer wishes particularly to express his appreciation to Mr. Philo T. Farnsworth and his brother, Mr. Lincoln B. Farnsworth, for tireless patience in explaining tne basic principles behind electronic television. Mr. A. H. Brolly, of Farnsworth Television Incorporated, was extremely kind in his help, as were Messrs. Richard L. Snyder, Frank J. Somers, and Lieutenant William C. Eddy, of the same company. Appreciation is also due Mr. W. Parmelee West, in charge of electrical communication at The Franklin Institute, for help and suggestions.

July 20, 1936 GEORGE H. ECKHARDT

INTRODUCTION

Electronic Television Breaks through the Laboratory Door

The life history of every important modern invention, in some respects, is very much similar to that of a child. First a child is carefully watched over and guarded by its parents, then comes a day when the child must take its first step—a step out into the world. The parents still watch over the child, guiding and counselling and giving help when help is needed. But the child then begins to make its own way in the world on its own merits.

This day of "emergence" has come for electronic television. Elaborate and costly research has been progressing in the laboratory. Now electronic television steps out into the field. It has ceased to be a thing about which engineers and scientists cautiously wrote in the transactions of learned societies, and about which few indeed outside of the laboratories where it was born knew anything definite. Up until now the people in the United States who have actually seen the present day electronic television receivers in operation could possibly be numbered in the hundreds.

Here is something new. A new science and a new art is springing up, offering opportunity in many fields. A new form of entertainment in the home, a new field of study for the radio amateur, and a new field for the technician and engineer is also born. The veil has been lifted, and electronic television makes its bow to the public and the engineer.

What Is the Present Status of Electronic Television?

There can be no doubt in the minds of even the most skeptical that television must eventually come in a commercially perfected form, simplified for everyday use, for as in the motion picture, sound followed vision, so in radio, vision must follow sound. In entertainment and study—in life itself—sound and vision normally go hand in hand.

Fig. 1. Mr. Philo T. Farnsworth (holding tube) and Dr. Rolf Moeller, of Fernseh, A.G. This tube is part of the research even now going forward to perfect a tube which will project the picture. The projection tube in the receiver is the next big step forward that can be expected in ELECTRONIC TELEVISION

Many attempts have been made in the past to bring out television, but most of these old attempts used mechanical devices which were complex and cumbersome even though the history of all inventions has shown that operations may be more speedily and accurately accomplished by electrical methods than by mechanical means. Electronic television has none of the disadvantages of older mechanical methods. It is, as its name implies—founded on electronic principles—made possible through electronic devices.

It would seem well to direct attention at this point to the background of the research that has made electronic television possible.

The basic research has been pursued largely, in fact almost entirely, in two laboratories in the United States. It has been costly research, and for this, and other reasons, there have not been great numbers of individual experimenters. The names of Philo T. Farnsworth, of Farnsworth Television Incorporated; and Dr. V. K. Zworykin, of the Radio Corporation of America, will always stand out as pioneers in the field of electronic television research.

Naturally the work has been carried on more or less secretly, and for this reason the general public has really never known the vast extent of the research. Both of these laboratories having ample financial resources, the work progressed with a minimum of publicity. From time to time dignified papers appeared in the transactions of engineering and other learned societies, and the research was conducted and reported in accordance with the highest engineering ethics. Thus the history of the research work in electronic television has been peculiarly free from the financial and other complications that usually hazard new developments, at least from the public's standpoint.

Europe Has Nothing in Electronic Television Which the United States Does Not Have

Despite the wide publicity that has been given television progress abroad, *neither England nor Germany have any-*

thing that the United States does not have. Farnsworth Television Incorporated, has agreements for the complete exchange of patents and technic with Baird Television, Ltd., of England, and Fernseh, A. G., of Germany; while the Radio Corporation of America has certain patent and license agreements with Electrical and Musical Industries, Ltd. (E. M. I.) of England, and Telefunken, of Germany. So

Fig. 2. Dr. Zworykin, director of the Electronic Research Laboratory of the R.C.A. Manufacturing Co.

far as practical developments are concerned, this embraces the electronic television situation of the world.

When Will the Public Have Electronic Television?

Naturally this is a question that is in the minds of the vast majority of people. And it is as yet unanswered because, to the great credit of the two pioneers in the field of electronic television, they have gone forward with extreme caution, and have made no fantastic promises or statements of any kind. The infancy of electronic television has been a healthy one in the laboratory. Its early childhood in the field promises to be equally healthy. Even now, it is being built on solid ground.

So in attempting to answer this question it is best to stick closely to actual facts. Two electronic television stations are now built, that of the Farnsworth Television Incorporated, in Philadelphia; and that of the Radio Corporation of America, in New York. These are, of course, considered experimental stations, but from them much needed data will be obtained—the kind of information that can only be obtained in the field. Electronic television has stepped out of the laboratory, it has crossed an important threshold. The first great step outside has been taken.

The present status of electronic television has been compared to radio in its "head-phone" stage. This statement is at the same time true and untrue. Rapid developments have been made in electronic television, and today it seems far ahead of the radio of "head-phones and crystal sets." The developers of electronic television have profited by the mistakes that were made in the introduction of radio, and have deliberately withheld it from the public until it has, at the present time, much higher entertainment value than did radio in its "head-phone" stage.

Transmission, however, offers a greater problem than confronted radio. Since electronic television must be transmitted over the ultra-high frequency bands (ultra-short waves) the effective radius of each station is not as great as that of a regular radio broadcasting station. This means that more stations will be required to cover a large area. Yet ten stations placed in key cities would serve a large proportion of the population of the United States.

The recent highly successful demonstration of the R.C.A. three-meter circuit between New York and Philadelphia, using two unattended and automatic relay stations, would seem to indicate that the problem of distance transmission over the ultra-high frequency parts of the radio spectrum has been successfully solved. While this circuit has not been officially tested for the transmission of television, yet there seems to be no good reason why it cannot be used for that purpose.

When it became expedient to "hook-up" chain broadcasting over land wires (what are now known as radio networks), the existing long distance telephone wires and cables were available and ready for use. But, special co-axial cables will be necessary to tie electronic television transmitting stations together for simultaneous transmission on a basis comparable to radio broadcasting. Fortunately, however, these same special co-axial cables will also be used for increasing the number of telephone conversations that can be carried, so in fact, electronic television is merely keeping apace with telephone research.

Electronic television right now offers a good and clear picture, with unusually high entertainment and educational value. This picture is shown in a cabinet that is about the size of the present-day console radio cabinet and can well be accommodated in the average living room. The unit is self-contained and there are no elaborate motors and moving parts.

Research work is already well advanced on a new tube which will make it possible to project an electronic television picture on a screen. This is mentioned merely so that it may be realized that research is going forward at a rapid pace.

The eye is a far more critical organ than the ear, and for that reason a far higher degree of perfection will doubtlessly be demanded of the electronic television receiver than was demanded of early radio. Again, from the visual standpoint, people have come to look upon the motion picture as a criterion. Now the motion picture is practically a perfected art. So it will be seen that the standards of comparison set for electronic television are far higher than those that were set for early radio.

The question, "When will there be electronic television in the home?" can only be answered by the people asking the question, paradoxical as that may seem. Despite the problems involved—the arranging of programs—the stringing of new lines and cables for station-to-station transmission—the actual building of stations, and other problems—there will be television in the home as soon as people want it—when they express a desire for it and demonstrate a willingness to pay for it. Electronic television was purposely kept in the laboratory until engineers were sure that it was ready for a field test. These engineers are now agreed that it is ready to emerge from the laboratory stage, and future problems will have to be worked out, just as the early problems of radio broadcasting were worked out, in actual field operations, in which a large number of people played an active part.

What Part May the Radio Amateur Expect To Play in Electronic Television?

The amateur has been a pioneer in the field of radio communication. He explored wave bands that, at the time, had not been particularly useful for other purposes, and then, when other uses were found for those bands, he was forced into bands of shorter and shorter waves. He now stands on the threshold of having the ultra-short wave bands more or less snatched away from him for the advent of electronic television.

For this reason, if for no other, Philo T. Farnsworth feels that the amateur should have a part in the further development in electronic television. To be sure, the electronic television receiving set will be a far more complicated thing to build, and will require a far greater skill and knowledge, than did the radio sets built by early amateurs. But it must be remembered that the amateur of today is a far more learned man than was the amateur of the early days of radio, and he will be able to accomplish far more than did his brother of fifteen years ago.

Parts will be available, especially the cathode ray receiving tubes, and there should be no serious obstacle in the path of the amateur and the radio experimenter wishing to plunge into electronic television. And it is to be hoped that the amateur will prove himself as valuable in the coming development of this new field, as he did in the field of radio development. Many of the present-day radio engineers have come from the ranks of the amateurs, and it does not seem too early to predict that electronic television will have to look to this source for many of the men who will be required in television work in the years ahead.

Electronic Television and Radio

Fear has sometimes been expressed as to the effect of electronic television upon radio. This fear seems to be groundless since it is not at all the purpose of electronic television to displace radio as it is known today. Present-day broadcasting will always have its place, and electronic television does not in any way encroach upon radio broadcasting any more than did sound motion pictures. In any event, radio broadcasting, and its many ramifications in the business and advertising fields, is safe for many years to come, no matter how great strides are made by television. It would be absurd to even think that a huge industry such as radio has grown to be could be scrapped over night, and it is not at all the purpose of electronic television to attempt this.

Electronic Television and the Motion Picture Industry

After careful consideration it seems that, in the long run, the motion picture industry has much to gain from electronic television. It seems safe to assume that, if there are no untoward restrictions made by the motion picture industry itself, a large part of the television programs will be made up of films. Sound motion pictures lend themselves very well to electronic television. The motion picture producer will eventually be able to sell his entertainment directly to the consumer in his own home; and will no longer be forced to become involved, in one way or the other, in the purchase and ownership of theaters in which to show films.

And again the motion picture studio technic—lighting, direction, etc., which have been so highly developed as arts by the cinema—will lend itself readily to the television studio.

* * *

Thus it will be seen that electronic television is not obstructed by any insurmountable barriers, either in the engineering or economic fields. Now that it has emerged from the laboratory into the field, and due to the fact that it was kept in the laboratory until engineers were certain that it was ready for field tests, it is reasonable to believe that capable minds will solve further problems as they present themselves.

It seems best to look upon electronic television as a new art, an art calling for a new technic all around—new actors, new directors, new technical men—and everything else new. It must not be looked upon as something that will take the place of anything now extant—it is new. New engineers, new research men, a new type of trained technical men, will be needed. Opportunities of many kinds will come to the fore. Therefore, instead of thinking that electronic television will displace this and that, and instead of comparing it with radio and motion pictures, it is to be regarded as the beginning of a new, and added, means of education, diversion, and entertainment; and it is to be looked upon as a wide new field for employment both in industry and the arts.

George H. Eckhardt

ILLUSTRATIONS

CONTENTS

PART I

THE PICKUP AND TRANSMISSION OF ELECTRONIC PICTURES

PART I

THE PICKUP AND TRANSMISSION OF ELECTRONIC TELEVISION PICTURES

CHAPTER I

FUNDAMENTALS

Electronic Television a New Field

Electronic television opens an entirely new field in the vast realm of radio communication, but it must not be assumed that it will displace the present broadcasting of sound, for that is not its purpose. It will most certainly, however, offer new opportunities for the amateur and radio engineer alike. Here will be found a new sphere, the surface of which has not been scratched.

The research and development of electronic television has, for the most part, been carried on in two private laboratories in the United States, and therefore very little regarding its inner working has been made public. What has been released from time to time has been more or less general. It is the purpose of this book to be, above all else, definite, and to give both the amateur and the engineer a firm grasp upon this subject, a subject destined to become more and more important.

Electronic television has little in common with the television that has preceded it, for in it there are no mechanical parts, there is no mechanical scanning, everything is done electrically.

Nor should electronic television be confused with **telephotography**, because it affords practically instantaneous transmission and reproduction of scenes with the added illusion of motion. Telephotography is neither instantaneous, nor does it give motion.

The whole subject of electronic television is best approached, it would seem, by asking and answering the two following questions:

1. What does electronic television seek to do?
2. How does electronic television accomplish this?

What Does Electronic Television Seek To Do?

This question is best answered by starting with a scene and following it through its various steps until it is reproduced at a receiver. First: the scene is transformed into an **optical image** by a lens, in the same manner that a camera, either still or motion picture, works. Second: this optical image is transformed into an **electron image.** Every point of this electron image gives off emissions which are in proportion to the amount of light falling upon that particular point. Third: this electron image is then taken apart in an orderly manner, in sequence, "dissected" as it were, and the emissions are transformed into electrical impulses, which, in turn, are proportional to the amount of light falling on the respective parts. It must be remembered that in electronic television this "taking apart" or "dissection" is done electrically; there are no mechanical parts. Fourth: the electrical impulses, or signals, after proper amplification, are transmitted over a single channel either through the air or over specially designed cables to a **receiver.** Fifth: in the receiver the procedure is reversed, and the "dissected" picture is reassembled.

This procedure, it will be seen, would give one picture, a "still," but **motion** is desired. In order to give the illusion of motion, and to transmit motion from a moving scene, the above procedure is duplicated many times a second, from 15 to 60. The number of times per second that this procedure is duplicated is known as **pictures** or **frames** per second.

This briefly outlines what electronic television seeks to do. The second question will not be gone into.

How Is This Accomplished?

While it is not the purpose of this book to go deeply into the theories of electronics that make electronic television possible, yet an understanding of the basic theories behind electronics must be understood in order to grasp the subject. These theories are briefly as follows:

The Electronic Theory

Scientists now almost generally agree that the atom, the smallest particle into which it was hitherto assumed that any element could be divided, itself is made up of a system of two kinds of even smaller particles; each individual atom consists of one **proton** with a positive charge of electricity and varying numbers of **electrons** with negative charges.

The proton, and some of the electrons, form a central "kernel," which is the smaller, but more massive, part of the atom. The remaining number of electrons in the atom circulate about this "kernel" much in the same manner that the planets revolve about the sun. The number of electrons and their arrangement about the proton are different in each element.

These roving electrons are free to move in their paths, but are restrained when they reach the boundary of the atom. If, however, energy is supplied to the roving electrons from outside of the atom, one or more of the roving electrons may approach the boundary of the atom with sufficient force to break loose.

The roving electrons may be compared to marbles rolling around in a deep dish, such as a soup plate. They can roll around the bottom but cannot escape. However, if some outside force is applied, they can be given enough impetus to roll up the edge and become free.

The Photoelectric Effect

With the electronic theory itself in mind, the photoelectric effect should be briefly gone into, because this is the very basis of electronic television.

It has been found that light falling upon certain metals gives the roving electrons of those metals sufficient energy to escape the boundaries of their atoms, and the **emission,** or escape, of these electrons, and their subsequent passage through a conductor, constitute the **photoelectric current.**

This, of course, is the theory upon which the **photoelectric cell** works, for the emissions, and subsequent photoelectric current, are proportional to the amoun[illegible] upon the cell. While the photoelectric cell is familiar to most readers, it may be well to treat upon it briefly, because of the great importance of the principle behind it in **electronic television.**

Several metals have been found to show outstanding photoelectric properties: lithium, potassium, rubidium, cae-

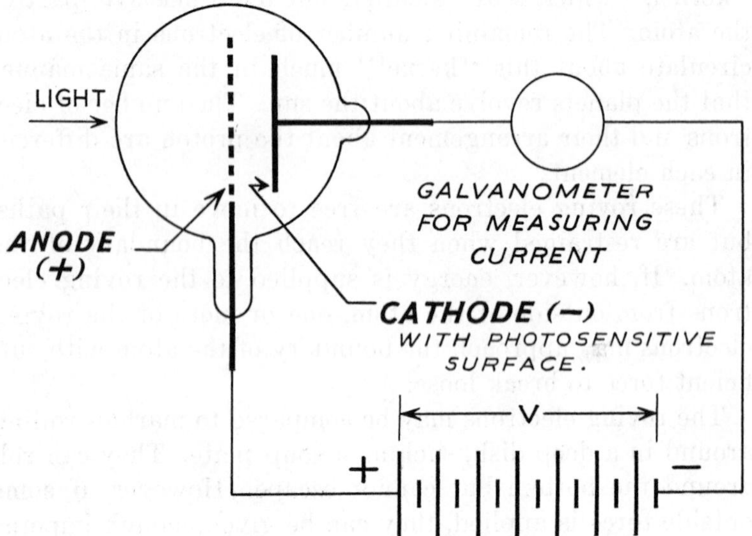

Fig. 3. Schematic Diagram of the Photoelectric Cell

sium, nickel, aluminum, sodium, strontium, and barium; for when these metals in special forms are subjected to a light source, they emit, or give off, electrons.

In short, a photoelectric cell consists essentially of a **cathode,** or negative part, composed of a thin film of photosensitive metal on a prepared base. This is connected, of course, with a negative potential.

There is a second metal electrode, the **anode,** or positive collector. This may be a ring of wire, or gauze, or a grid of wires connected with a positive source.

Essentially the photoelectric cell consists of a cathode with a photosensitive surface, and an anode to collect the ph[illegible]ted when light falls on the cathode (see Fig. 3). When this cathode is illuminated, photoelectrons are emitted, and the emissions are in proportion to the intensity of the light. The cell is connected, as shown, with a battery of sufficient voltage so that the emitted electrons from the cathode may be "drawn off."

As the intensity of the light varies, the readings of the galvanometer, measuring the photoelectric current, will vary.

The photosensitive elements are exceedingly active chemically, readily combining with oxygen. So in practice they are inclosed in a tube containing an inert gas, or vacuum.

When the cathode of a photoelectric cell is exposed to a light source, there are electron emissions from it, and these emissions are in proportion to the varying intensities of that light source.

Scanning

With these fundamental principles in mind, the reader is ready to approach the important subject of **scanning**.

Taking the picture in Fig. 4 as the "picture field," or "optical image," it at once becomes clear that nothing would be gained in exposing the entire picture to the cathode of a photoelectric cell. Yet each section of the picture reflects an amount of light in proportion to the light and shadows of the various points on the picture. The human eye sees the picture because it is able to "dissect" the optical image into small parts and ascertain the proportional amount of light coming from each part; and then assemble these "signals" as a picture again in the "receiver," the brain.

Suppose that instead of exposing the whole illuminated picture to the cathode of a photoelectric cell, a small square aperture h units on a side, that is, h square in area, is moved across the picture in closely adjacent parallel strips.

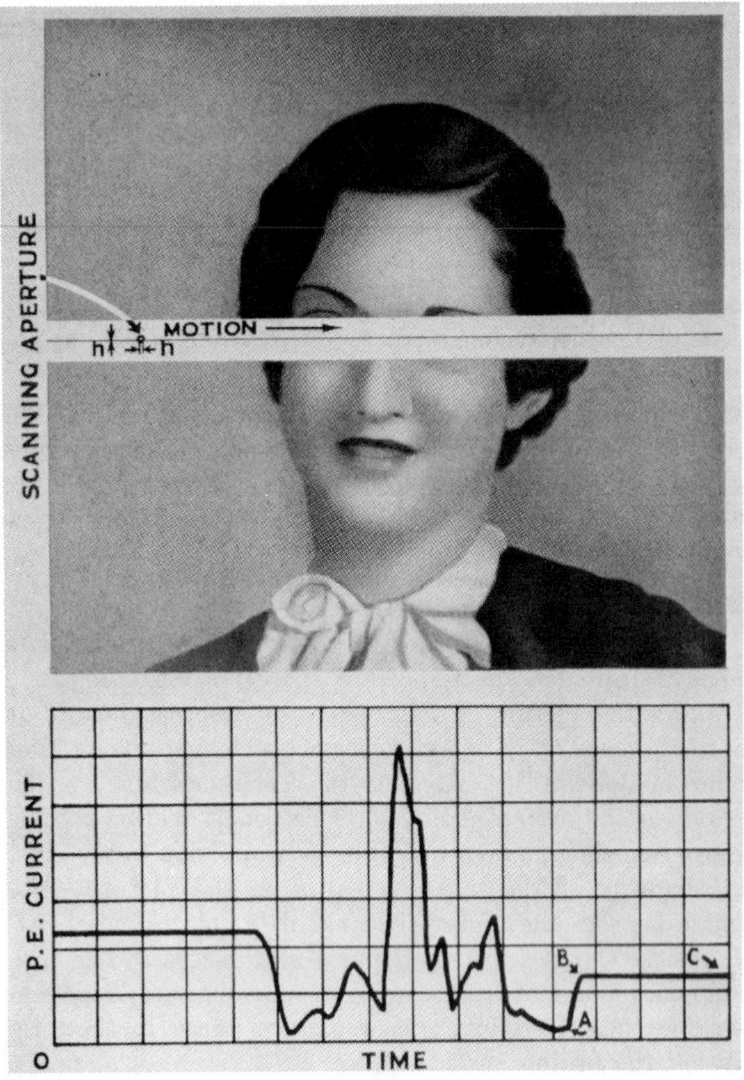

Fig. 4. The Fundamentals behind Scanning. Graph shows varying photoelectric current as aperture passes across picture

The light coming through this moving aperture is exposed to the cathode of a photoelectric cell. It will be seen that the photoelectric current resulting will vary with the light and shadows of the picture. Where the picture is of one intensity, as from B to C, the photoelectric current remains constant, a straight line on the graph.

If the whole picture is H units high, the number of times that the aperture will pass over it to cover the entire picture will be

$$\frac{H}{h} = n.$$

1. It will thus be seen that the picture has been "dissected" in a number of little squares, h on a side, and the different average intensities of light in each square have been converted into different respective intensities of electric current.

2. It will also be seen that the height and breadth of the picture, two dimensions, have been reduced to a single continuous line. In short, the picture has been cut into a number of strips, and these strips are put together in one continuous strip.

Thus, from 1 and 2, it will be seen that the two characteristics of an electric current, amplitude or magnitude and duration of time, are achieved, and it is possible to convey the light intensities and space dimensions of a picture over a single channel.

The receiver merely reverses this process of **dissection** and reassembles the picture.

CHAPTER II

THE FARNSWORTH SYSTEM

With these basic principles in mind, the reader is ready to follow in detail the design and workings of electronic television. It would seem best to first take up the matter of **pick-up**, following the picture through the entire process until it finally again appears as an optical image at the receiver.

The Pick-up or Image Dissector

The system of electronic television for which Philo T. Farnsworth is responsible is extremely ingenious, but also quite simple considering the problems involved. Each of these problems has been met after years of untiring research.

In the **Farnsworth System,** the transmitter centers about the pick-up, which is known as the **Image Dissector.** Briefly, this is a vacuum tube which converts the various light intensities of a scene focused upon its photosensitive surface into fluctuations of an electric current. In addition to this it "analyzes" the area of the scene into a regular succession of space elements, converting them into corresponding signal currents that can be transmitted over a single communication channel.

The Farnsworth Image Dissector will best be understood if its "evolution," as it were, is studied, taking up each difficulty encountered, and explaining how it was met.

Essentially the image dissector tube is an evacuated tube, with a silver oxide-caesium cathode, in the form of a disk. This disk is perfectly smooth, almost polished, and gives the photoelectric element required.

The Evolution of the Image Dissector Tube

The steps in this evolution and the problems encountered and met are as follows (Fig. 5A and B):

In Fig. 5A, the scene which is being **televised** is sharply focused upon the photoelectric sensitive cathode by means of a lens, just as in a camera the scene is focused upon a plate. In the ordinary camera, when the scene is focused upon the plate, the varying intensities in light in the picture

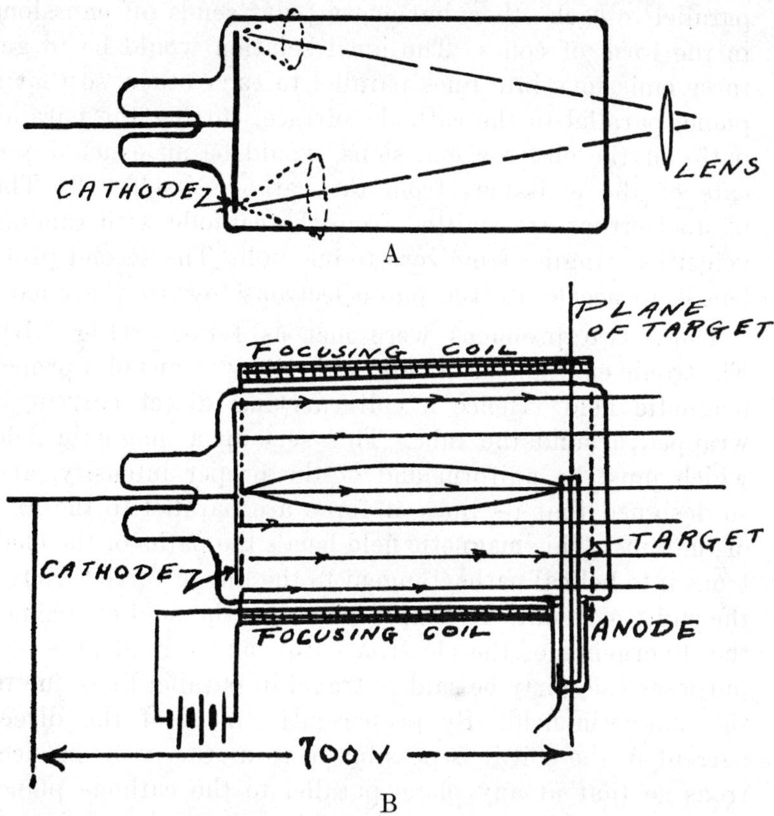

Fig. 5. A & B, Evolution of the Farnsworth Image Dissector

cause varying chemical changes in the emulsion on the plate. This, of course, is the principle behind photography. In the image dissector tube, however, the varying intensities of light in the picture cause varying **electron emissions** from every point on the surface of the photoelectric sensitive cathode. The cathode surface in the image dissector tube being polished, it is comparatively easy to focus the scene sharply upon it by means of the lens.

Two problems at once present themselves: (1) The electrons or photoelectron emissions from every point on the photosensitive surface do not come off in straight lines parallel to each other, but every point sends off emissions in the form of cones. The ideal solution would be to get these emissions into lines parallel to each other, so that a plane, parallel to the cathode surface, cutting the parallel paths of the electron emissions, would be an exact duplicate of the emissions from the cathode itself. (2) The photoelectrons are emitted from the cathode with random velocities ranging from zero to one volt. The second problem is to accelerate the photoelectrons toward the anode.

These two problems were met as follows (Fig. 5B): Electronic emissions can be "guided" by means of a proper magnetic field. Hence a coil carrying **direct current** is wrapped around the tube. This sets up a magnetic field which must be uniform and of the proper intensity, and so designed that its lines of force are parallel to the axis of the tube. This **magnetic field** bends the paths of the electrons into **helical paths,** tangent to the line of force through the emitting point. In simple language, the field overcomes the divergence of the electron paths, and for all practical purposes they may be said to travel in parallel lines due to the magnetic field. By proper adjustment of the direct current in the coil it is possible to focus the paths of electrons so that at any plane parallel to the cathode plane, there is a sharp focus of the emissions, giving an exact duplicate of the cathode emissions, point for point.

An anode is added, part of which is a "finger-like" arrangement embodying on apertured target. An accelerating voltage of approximately 700 volts is added to accelerate the photoelectrons.

It will be seen that the electron emissions of the cathode surface are traveling toward the anode, and that each of these emissions is proportional to the respective amount of light falling upon the cathode surface from the optical image being "picked-up." It will also be seen that these emissions can be sharply focused so that the emissions from the cathode surface are duplicated in the **plane of the target.** This second set of emissions, exactly duplicating those from the cathode surface form what is known as the **electron image.**

If this **discharge** — the electron image — is bombarded against a **fluorescent screen,** the optical image is reproduced. The fluorescent screen serves as a **transducer,** absorbing electrical energy and emitting light.

Having obtained an electron image, in the image dissector tube, the next problem is **How Can This Electron Image Be Scanned?** How can every point be gone over in sequence, taking off the various electronic impulses or signals in an orderly manner so that they may be transmitted and reassembled in a receiver?

Scanning Motion Picture Films

Assume that it were possible to have a small scanning aperture, or opening, on the end of a device, shaped like a lead pencil and forming part of an anode. Then assume that this device were moved back and forth across the electron image, from left to right, at a uniform rate, and then almost instantly returned to the left, and again moved across uniformly, along a line parallel and next to the line traced above. As this device moved along from left to right, electron emissions from respective parts of the electron image would enter the aperture and give a continuous signal. If this could be duplicated at a receiver, the image could be duplicated. Obviously all this would be impractical.

Now the paths of electrons can be deflected by magnetic fields. This fact offers a solution to the above problem.

Instead of moving the scanning aperture about, the scanning aperture is fixed, and the electron image is moved. The result is the same (see Fig. 6).

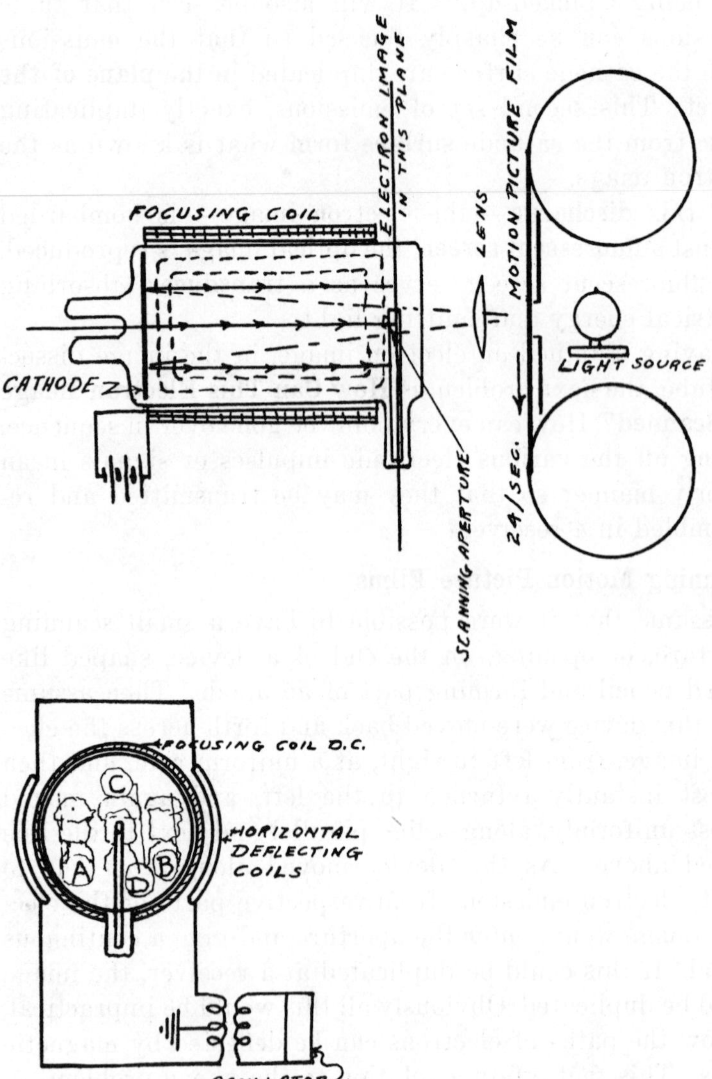

Fig. 6. The Farnsworth Image Dissector for Picking-up Motion Picture Film

Two motions were found necessary to carry every part of the electron image across the scanning aperture so that each part might be picked-up in an orderly manner in sequence.

The film was run past the opening at the rate of 24 pictures per second. It will be seen that the electron image would have the identical motion as the film. This, of course, gives the vertical motion.

The two coils, in series, and fed alternating current were added as shown. These coils deflected the electron image back and forth, from *A* to *B,* and *B* to *A,* as shown, while the motion of the film itself carried the electron image resulting from the optical image from *C* to *D.*

It must be noted that in this device the motion picture film moves steadily, and is not "jerked" from picture to picture, as in a projector.

While there are 24 motions from *C* to *D* per second, there are some 5,760 motions from *A* to *B* and back to *A* per second.

While the film is moving vertically at the rate of 24 pictures per second, steadily and uniformly, the horizontal scanning coils are adjusted to move the electron image back and forth across the scanning aperture at 5,760 times per second. Obviously it would be desirable to scan horizontally in only one direction, as from left to right, in order to approximate a straight line. Therefore the coils are adjusted to give a "quick return," and in this way a straight, continuous line is closely approximated. Thus, the electron image is comparatively slowly moved from left to right across the scanning aperture, and quickly returned to begin again.

The size of the scanning aperture itself must be worked out from the number of **horizontal cycles** or trips per second back and forth of the electron image. The number of horizontal cycles across each individual picture frame is known as the **lines** in television. Thus, if there are 240 cycles of horizontal scanning for every frame, the result is

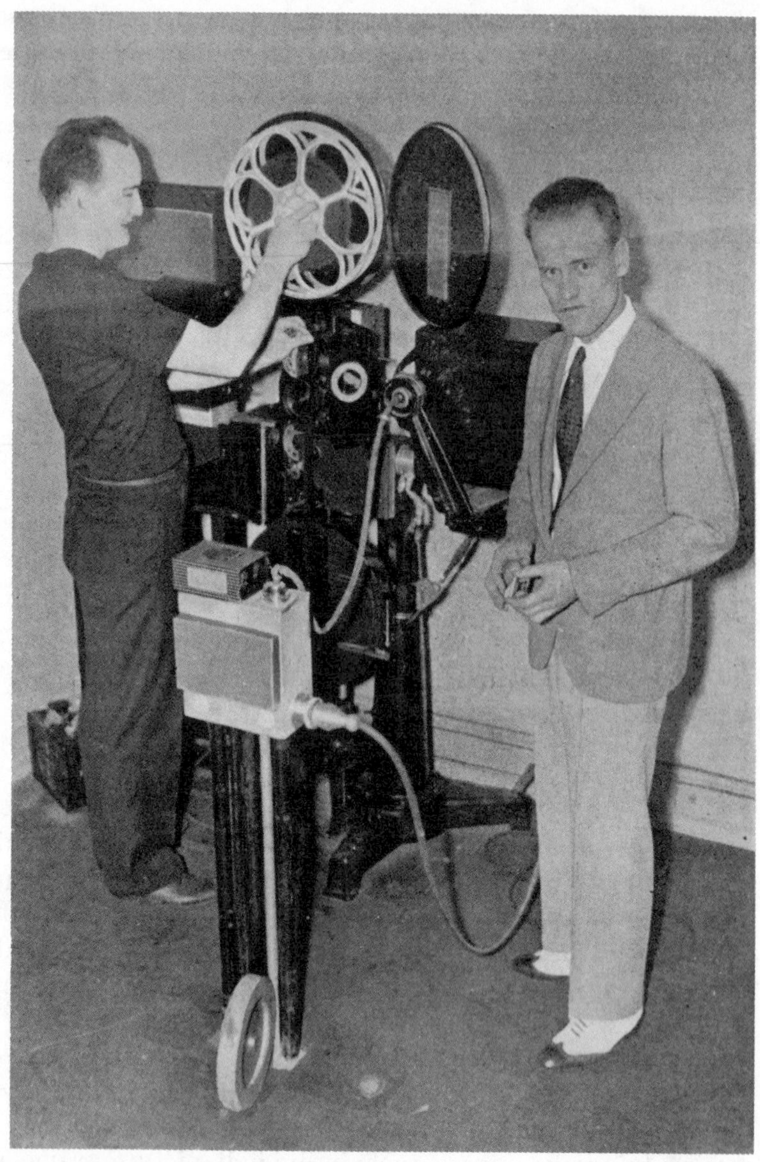

Fig. 7. The Farnsworth Telecine Used for Scanning and Transmitting Motion Picture Film

said to be **240 lines** (definition). If there are **240 lines** per frame, and **24 pictures,** or **frames,** per second, there are 5,760 horizontal cycles per second (24 x 240).

The matter of lines and pictures, or frames, in television, is of great importance, and will be gone into fully later. Now the electron image emitted from the cathode in the Farnsworth Image Dissector Tube under discussion is actually 4 inches in diameter. If 240 lines are desired, it is obvious that each strip to be covered by the scanning aperture should be $\frac{4}{240}$ or .017 of an inch wide. Actually the scanning aperture is made .015 inches wide.

The Telecine

The Farnsworth device for picking-up, or scanning, motion picture film is known as the **Telecine Projector.** The entire scanning and transmitting unit is very compact (Fig. 7).

The high development of the projector for picking-up motion picture films has brought about an exceedingly interesting technic in television pick-up, especially in Germany, where Fernseh A. G. has an agreement for the complete interchange of patents with Farnsworth Television Incorporated. This is known as the **Intermediate Film Process.**

The Intermediate Film Process

Briefly, the scene to be picked-up is "taken" on an ordinary motion picture camera, special film being used. This film immediately passes through a developer, fixer, and washing bath, and is partly dried, and then passed to a Telecine Projector, where it is scanned. The process has been perfected to the point where there is a delay of only about one minute from the time the scene is being taken until the film is being scanned and transmitted.

The film is then thoroughly dried and stored. Fernseh A. G. has trucks equipped to travel about and pick-up scenes. These trucks are also equipped to transmit the scanned scenes.

Since there is a delay of something like one minute between the time that the camera "takes" the scene, and its

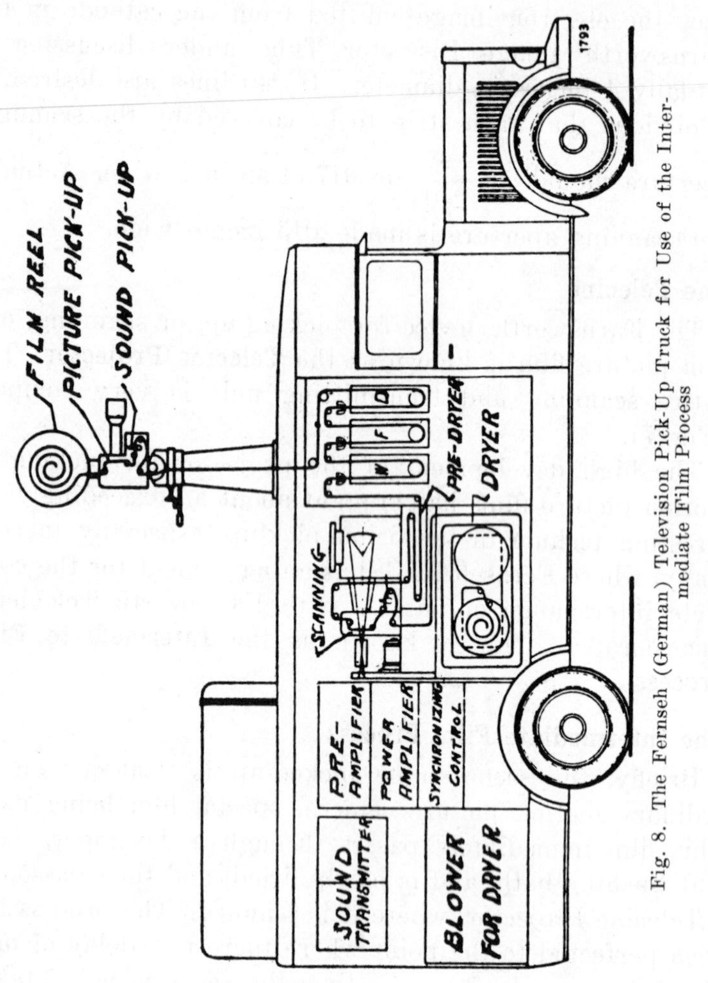

Fig. 8. The Fernseh (German) Television Pick-Up Truck for Use of the Intermediate Film Process

subsequent scanning and transmission, the film used is of the sound type, and the sound transmission is from the sound track on the film (see Fig. 8). In this truck the exposed film goes through developer (*D*), fixing bath (*F*), and washing bath (*W*). It is then partly dried, scanned, and then fully dried and stored.

The Farnsworth Multiplier

The evolution of the Farnsworth Image Dissector has been followed to the point where a dissector is required that will scan actual scenes both in the studio and outdoors.

Two new problems then presented themselves: (1) In the scanning of motion picture film, the light could be controlled so that it would always approach the ideal. Now in the studio and out-of-doors this control of intensity of light could not be so well ordered; in fact, out-of-doors light must usually be taken as found. The result of this problem was that often the impulses from the electronic emissions caused light falling on the cathode were extremely feeble, and some new means of amplification was necessary. (2) In scanning motion picture film the path of the film past the lens, at a steady rate of speed, supplied the vertical motion necessary for scanning. However, in scanning actual scenes direct, this vertical motion had to be supplied.

This second problem was solved by placing another set of coils, in series, and fed alternating current, above and below the tube. These coils deflect the electron image up and down, while the first set of coils (horizontal) deflect the image back and forth. Thus this second set of coils provide a motion to the electron image similar to the motion supplied by the film in the Telecine.

The vertical coils have a frequency of from 15 to 60, that is, they are capable of scanning from 15 to 60 electron image pictures per second. Thus they are known as **low frequency coils.** These coils have from 5,000 to 10,000 turns each.

The first problem has been most ingeniously solved by the **Farnsworth Multiplier,** which is incorporated as a part of the Farnsworth Image Dissector Tube.

CHAPTER III

SECONDARY ELECTRON MULTIPLICATION

Before taking up the Farnsworth Image Dissector Tube in its complete form, it is necessary that the reader have an insight into **secondary electron multiplication,** a subject important not only in television, but in the future of the whole science of radio communication.

Amplification

The science of radio communication has been built up largely because of the availability of devices which will amplify feeble and very rapid electrical variations. In short, amplification is a very important factor of radio communication.

The amplifiers in use are all essentially relay devices in which a feeble electrical voltage "triggers off" a constant source of power in such a manner as to give a new electrical variation similar in all respects to the original except of much greater power. This process is repeated successively many times until the final variations may be more than a million times greater than the original electrical impulse.

The extent to which such amplification may be carried, however, is limited. This is because electric charges are not a homogeneous fluid, but have a definite atomistic structure. To use an analogy to explain this, electric charges are not like water flowing, but rather like a stream made up of extremely small shot.

Two forms of interference arise because of this fact. The first of these is directly due to the corpuscular "grain" of the electric fluid. This interference, or "noise" as it is called, is produced by the grain size of the current, and is called the **Schotke effect,** and may be likened to the noise produced by the patter of rain upon a tin roof.

It is a matter of common observation that the amount of

noise produced by rain increases as the rainfall becomes heavier. Similarly, the amount of fluctuation noise generated in an amplifier is proportional to the electrical current which is used in an amplifier. In the ordinary hot cathode tube type of amplifier, widely used in radio today, the total current flowing across the tube may be a million times larger than the component of that current which represents the amplifier signal.

Another source of interference which limits the amount of electrical amplification is known as **thermal noise.** This is due to the fact that the electrons in a substance share the movements of the molecules in the material, and thereby produce rapidly varying electric currents in the elements of the amplifier. This results in random voltages being applied to the input of the amplifier, which cannot be distinguished from signal impulses of the same order of magnitude.

In television these small effects become highly important, since they limit the amount of amplification to be used in the image pick-up device. There are two reasons for this: (1) The electric currents generated by the transmitting device are extremely feeble. (2) The duration of certain components in the picture currents are so short that as low as 5 or 10 electrons may represent the total quantity of electric charge involved.

To overcome these difficulties Philo T. Farnsworth developed a system of electron multiplication for amplification to be used in the Farnsworth Image Dissector Tube. This furnishes amplification with a much lower "noise" level than can be obtained with the ordinary thermionic ("hot tube") relays.

The Farnsworth Multipactor Tube

Secondary electron multiplication can best be understood by going back to the original analogy of the electrons of a substance being likened to marbles, or shot, rolling around in a soup plate. If a single marble, or shot, is thrown into this plate with considerable force from the outside, it will "splash out" a number of the shot in the plate. If energy

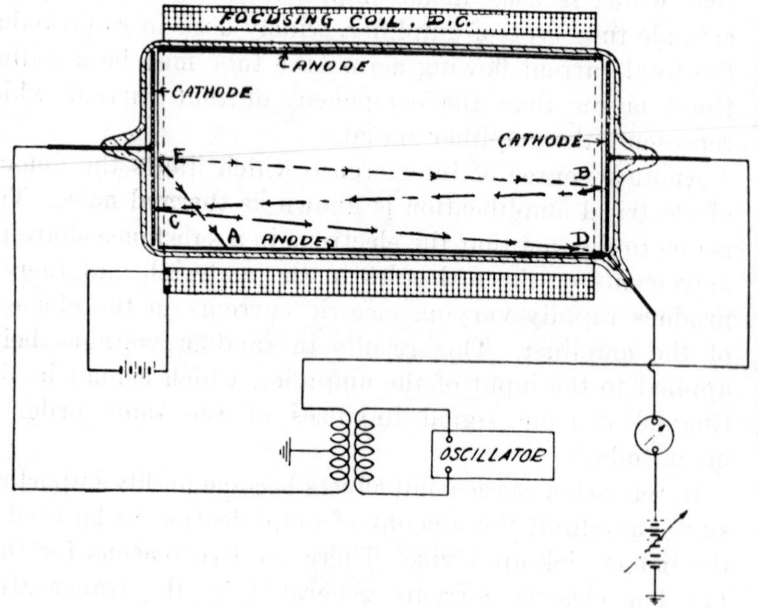

Fig. 9. Radio Frequency Type Farnsworth Multiplier

were supplied to each of these "splashed-out" shot, and each of them, in turn, were hurled into other plates, more would be "splashed out." In this way, as the process went on, more and more shot would be splashed out and these could eventually be gathered by a centrally located plate.

This is exactly what happens in the **Farnsworth multipactor tube**, so called because the process is the result of multiple electronic impacts (Figs. 9 and 10).

Essentially a multipactor tube consists of a cylindrical evacuated envelope, having plate like cathodes mounted in each end. These cathode plates are made from pure silver, oxidized, and formed with caesium into sensitive secondary emitting surfaces. A cylindrical anode of nickel or molybdenum is positioned between the cathode plates. This anode may fill almost the entire space between the cathodes, or may be merely a ring.

A coil surrounds the envelope in the form of a solenoid.

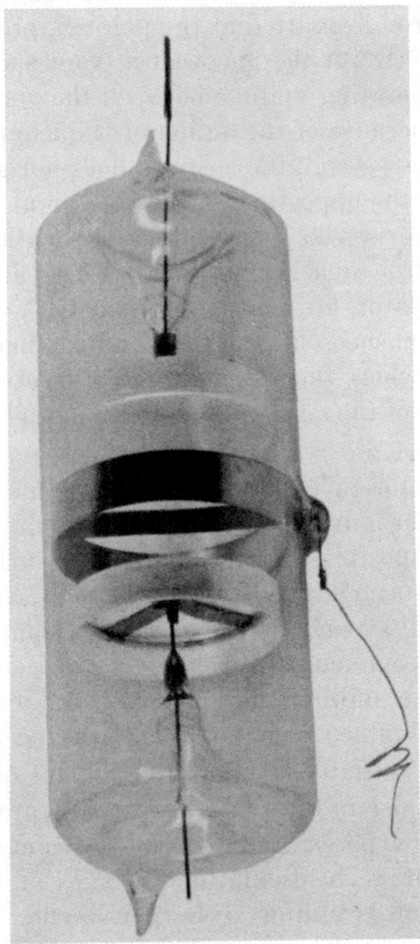

Fig. 10. Farnsworth Radio Frequency Type Electron Multiplier Tube such as outlined in Figure 9

This coil is supplied with direct current, which establishes a longitudinal magnetic focusing field between the cathode plates.

The cathode plates are supplied with radio frequency at

a frequency of approximately 50 megacycles, or higher. The anode is connected through a meter to the positive terminal of a variable potential source.

Briefly, the multipactor works as follows (see Fig. 9): If an electron is emitted at the point *E*, ordinarily it would strike directly for the anode at a point such as *A*, because of the accelerating voltage between the anode and cathode. However, because of the action of the magnetic field set up by the focusing coil, the electron does not go directly to *A*, but strikes the opposite cathode at a point such as *B*. Now the oscillator works between the two cathodes. At *B* the single electron smashes out more. These, in turn, by virtue of the oscillator, are sent back towards *C*, where more electrons are smashed out, and these in turn finally go to a point such as *D*, where the accumulation of electrons, many times the number of the single electron, are gathered by the anode.

Because of the extremely high multiplications obtainable in multipactor tubes, the tubes may be used as a source of electrons for purposes other than amplification. For instance, composite silver oxide-caesium surfaces, similar to those used in making photoelectric cells, have an emission of between 10^{-12} and 10^{-14} amperes per square centimeter at ordinary room temperature. If these feeble currents are multiplied a million million times, an ampere electronic output is obtained from cold metal surfaces.

Currents initiated by this process and of this order are in constant use in the Farnsworth Laboratories, being used for many purposes, particularly the conversion of direct current voltage to oscillating voltages of from 100,000 to several hundred million cycles per second. Such oscillators start by merely closing a switch in the battery circuit. It is not necessary to wait until they "heat up," as in the case of thermionic tubes.

The practical advantages of these multipactor tubes used as oscillators are their simplicity and very high conversion efficiency, which may be as high as 95 per cent. The very great advantage to be gained by this high efficiency is that, for a given power output, the tubes may be much smaller than corresponding thermionic tubes.

Electron multipliers have been made to perform every function now performed by the thermionic relay. While it is improbable that all of the functions now performed by thermionic tubes will be replaced by these new cold cathode multipliers; nevertheless, it seems evident that this new art of secondary electron multiplication will have a very revolutionary effect upon the science of radio communication.

These multipactor tubes have been a by-product, as it were, of Farnsworth television research, since they were largely developed to give ''noiseless'' amplification in the image dissector tube.

CHAPTER IV

THE FARNSWORTH PICK-UP CAMERA

Philo T. Farnsworth has combined the image dissector tube and the electron multiplier so that the sensitivity of the former may be increased. This combined tube is the "heart" of the pick-up camera used in the Farnsworth system of television.

Description

The photoelectric cathode, upon which the optical image is focused, is formed as a thin translucent film of silver oxide on one end of the glass tube itself. The multiplier is at the opposite end of the tube.

The main anode of the dissector tube is a silver disk having the edge spun over so that it fits snugly into the glass envelope.

Behind the main anode is a tightly fitting silver cup in the center of which the scanning aperture is punctured. This silver cup forms one of the multiplier cathodes. The second multiplier cathode is mounted from a stem. The output of the multiplier is taken from an anode ring which is supported from the tube walls.

After the tube has been mechanically constructed, it is sealed onto a pump and given a baking until a very good vacuum is obtained. Oxygen is then admitted and the dissector and multiplier cathodes properly oxidized into photo surfaces. Caesium pills are then flashed in both the dissector and multiplier chambers to give the proper amount of caesium in each, after which the tube is given a heat treatment to sensitize these cathodes.

Operation

The operation of the Farnsworth image dissector is as follows (Fig. 11):

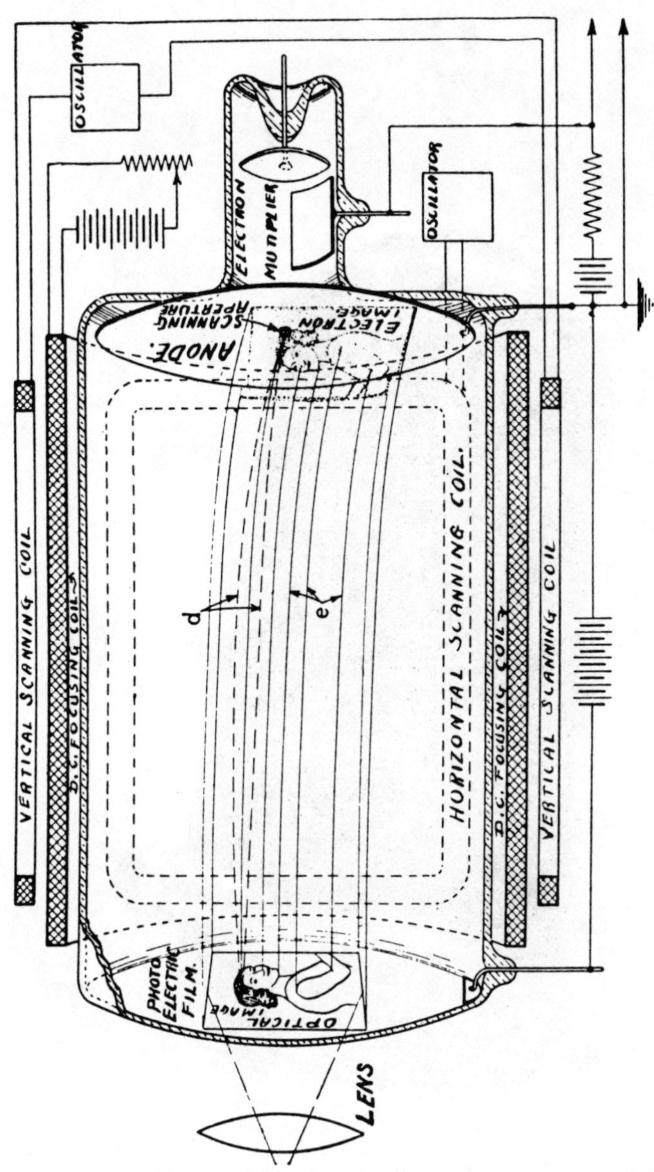

Fig. 11. Outline of Farnsworth Image Dissector Tube

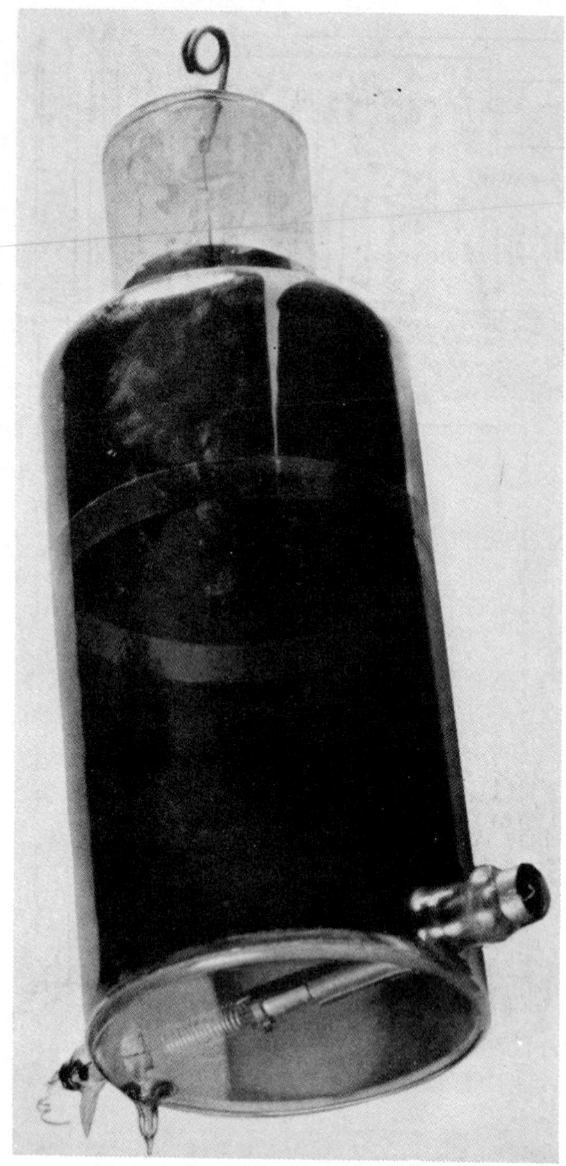

Fig. 12. Farnsworth Image Dissector Tube with Multiplier

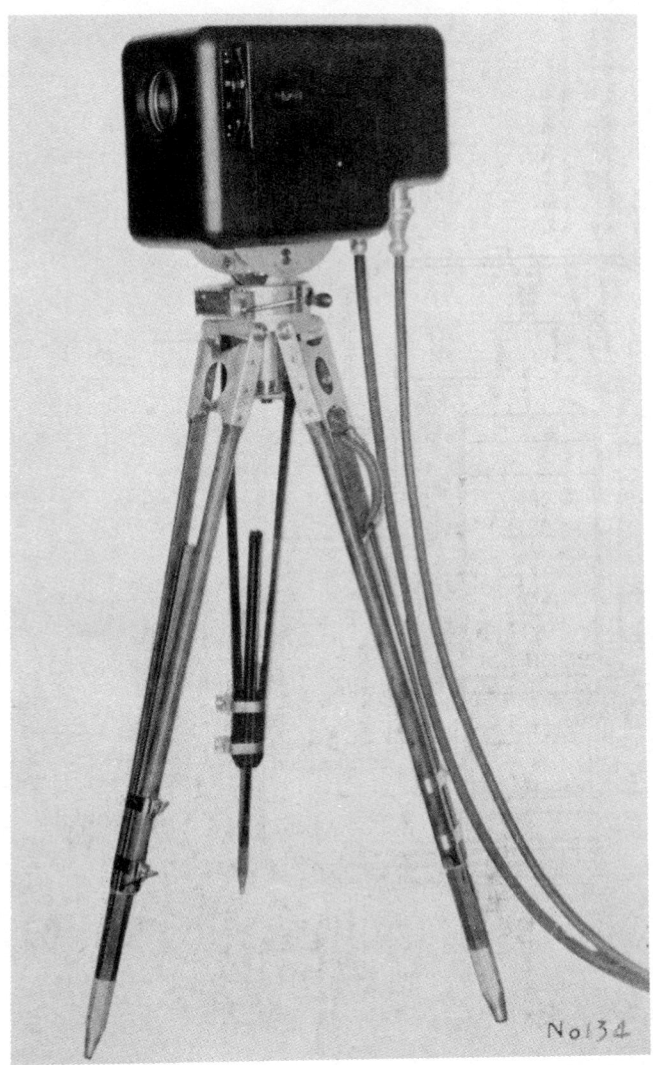

Fig. 13. The Farnsworth Pick-Up Camera complete. (Note compactness)

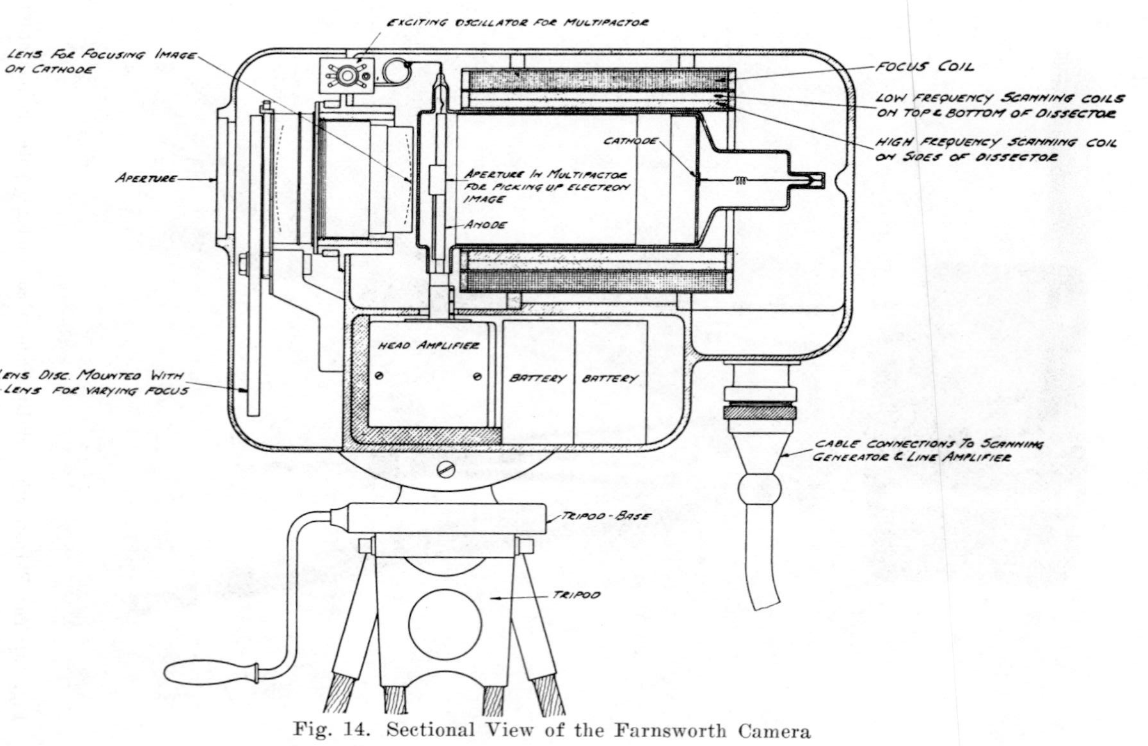

Fig. 14. Sectional View of the Farnsworth Camera

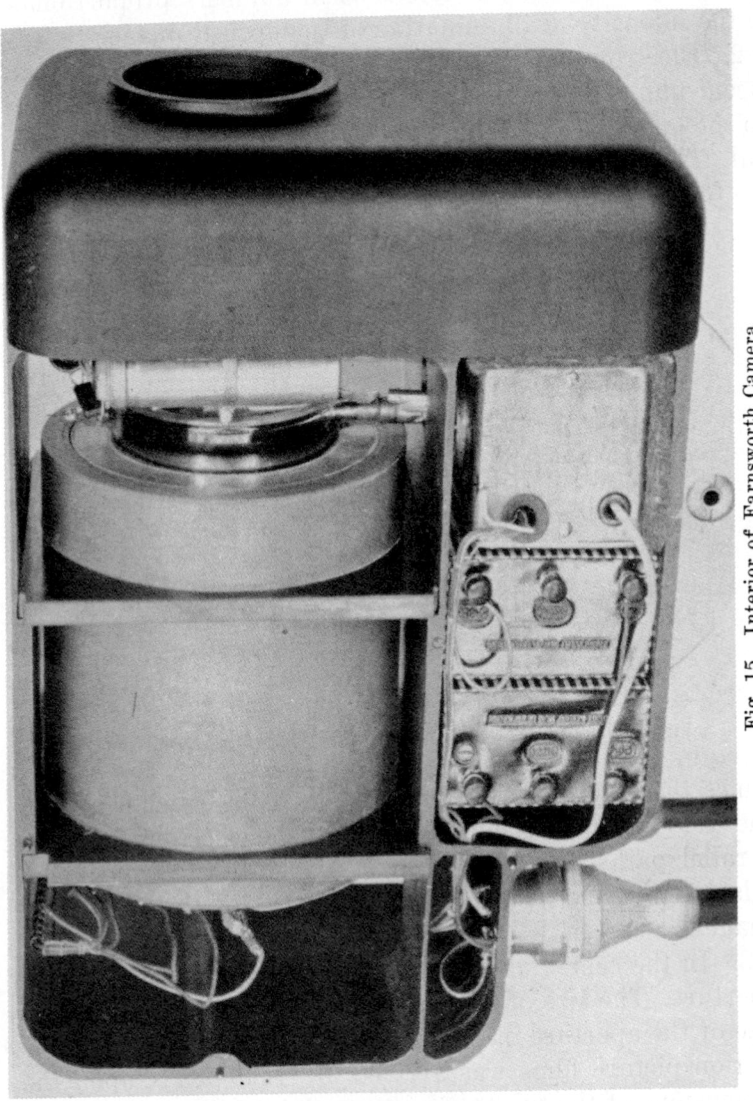

Fig. 15. Interior of Farnsworth Camera

1. A lens focuses an optical image of the scene to be transmitted on the translucent photoelectric film. This photoelectric film emits electrons in numbers proportional to the intensity of illumination of the optical image.

2. Electrons from any point of the image on the translucent film are brought to a focus at a corresponding point on the anode to which they are drawn by its positive potential. This focusing is accomplished electromagnetically by the focusing coil which carries D.C. current. This coil overcomes the divergence of the electron paths as shown at *d*, and the result is the same as if all of the electrons followed parallel paths as indicated at *e*. Thus an electron image exists at the anode, and this represents exactly the optical image.

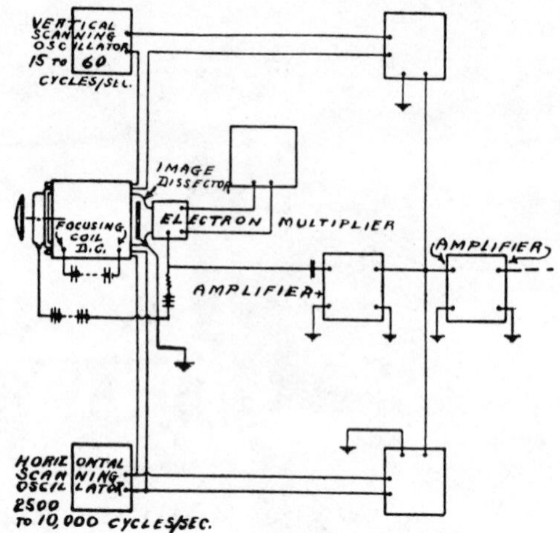

Fig. 16. Diagram of the Farnsworth System of Transmission

3. In the center of the anode there is a square scanning aperture. The size of this aperture is arrived at just as the size of the aperture in the target of the image dissector for motion picture films was arrived at. It depends upon the dimensions of the tube and the "number of lines" desired.

4. To scan the electron image, alternating current is fed to the horizontal and vertical coils shown. These coils deflect

the electron image back and forth, and up and down, past the scanning aperture. As a respective part of the electron image passes the scanning aperture, the electrons from that part pass through the aperture into the multiplier, from which they are gathered. Thus a constantly changing signal is obtained which represents the amount of light of a respective part of the electron image, and hence the optical image.

The frequency applied to the vertical scanning coils is from 15 to 60 cycles per second, while that applied to the horizontal coils is from 2500 to 10,000 per second.

Another definition of the number of lines, that important term in electronic television, is the horizontal frequency divided by the vertical frequency. Hence

$$\text{Lines} = \frac{\text{Horizontal Frequency}}{\text{Vertical Frequency}}$$

Thus, for a 240-line picture, repeated 24 times per second, the horizontal frequency would be 5,760 cycles.

The Farnsworth image dissector tube is easily capable of up to and above 500 lines.

CHAPTER V

THE R. C. A. SYSTEM

The Iconoscope

The **iconscope** is the vital center about which the **R. C. A. system of electronic television** is built. This pick-up device, which approximates the human eye itself, resulted from the experiments of Dr. V. K. Zworykin. The word iconoscope was taken from the Greek word *icon,* meaning "image," and "scope," signifying "observation." It has also been called the electric eye.

There are two outstanding, distinguishing features in the iconoscope:

1. The optical image is not focused upon a single photosensitive cathode, but upon vast numbers of separate and distinct minute photoelectric elements, each acting like an individual photoelectric cell, each registering the proportionate amount of light falling upon it from the respective part of the optical image.

2. The signals from each of these cells, each signal going out in sequence on a single channel and registering the amount of light falling on that cell, are obtained by a most ingenious discharge method.

It was necessary to surmount tremendous obstacles in order to make this device practical. In the first place, it is evident that the number of distinct photoelectric elements must be equal to, or greater than, the number of elements into which the picture is subdivided. Therefore, if a 243-line picture is to be transmitted, at least 80,000 elements will be required, while a 350-line picture will require some 150,000 elements. Furthermore, in order to produce the illusion of continuous motion in the picture, at the receiver, 20 to 30 complete pictures must be transmitted each second, so that the time available for transmitting the information from any one element will be only a small part of a microsecond.

Comparing this problem to that experienced in the case of photographic plates, the difficulty will be fully realized. In the photographic plate all of the points on its surface are affected by light during the entire period of exposure, which varies from several seconds in a studio to around one-hundredth of a second out-of-doors. In any event the time is many thousand times greater than in the case of the televised picture. The human eye operates under even more favorable conditions than the photographic plate.

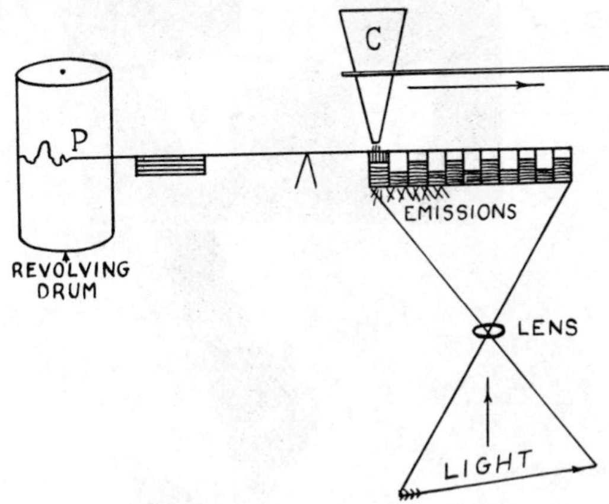

Fig. 17. Simple Explanation of Discharge Process

Therefore it was necessary to devise some method for storing the energy in these individual cells between two successive scannings. This was accomplished by placing the **mosaic** of countless individual photoelectric cells upon a backing of **mica**, which operated as a condenser, as it were, storing up the energy between scannings.

Before taking up the discharge process, it may be well to return to the basic principles of electronics.

Assume, as in Fig. 17, that on the right-hand side of a balance there exists a row of small cells filled with shot. These cells may each be likened to one of the many individ-

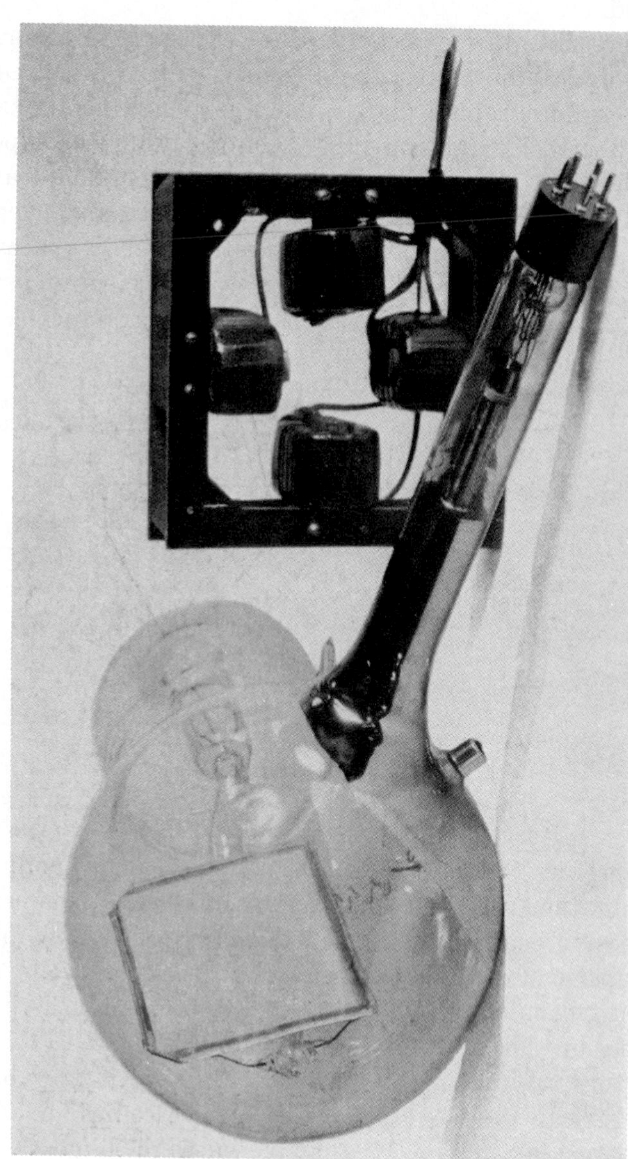

Fig. 18. The Iconoscope tube—the Vital Center about which the R.C.A. System Is Built

ual photoelectric cells, and the group would be analogous to a row of these cells. The shot might be likened to the electrons. This row of cells, each charged to capacity with shot (electrons), is balanced by a weight on the left-hand beam. A strip of light from an optical image is focused on the row of cells. According to the theory of photoelectricity, each cell will emit electrons in proportion to the amount of light falling upon that cell. Some will emit more than others. Naturally the balance will be upset.

However, a nozzle *C* moves across the top of the row of cells at a constant rate, filling each with shot, the amount poured in being the exact amount that has been emitted from each cell, the surplus shot from *C* overflowing.

Now as *C* fills each cell the pointer *P* will move, and the motion of *P* will be in proportion to the amount of shot (electrons) necessary to make up for the emissions. If a pen were placed at *P*, this would mark a curve which would show the amount of the emissions. This, roughly, is the analogy of the way the scanning signal is taken off the iconoscope.

Actually the iconoscope (see Fig. 18) is 18 inches long and the sphere is 8 inches in diameter. It is an exceedingly highly evacuated tube, containing:

1. The **mosaic.** This is the photoelectric element of the iconoscope. In actual practice the number of photoelectric elements on the mosaic is many times the number of elements making up the picture transmitted.

2. The **signal plate.** This is formed by a metallic coating on one side of a thin sheet of mica. The mosaic and signal plate form the **retina** of the inconoscope proper.

The mosaic can be produced by a number of methods, the simplest of which is the direct evaporation of the photoelectric element on the mica sheet in a vacuum. Another suggested method would be to rule the mosaic from a continuous film of photoelectric metal on mica, using a ruling machine.

The mosaic used in the iconoscope is [illegible] of silver globules each of which is photosensitized [illegible]esium.

3. In the neck of the iconoscope is [illegible]n **electron gun**. This, for convenience, is placed at an angle of 30 degrees to the normal passing through the middle of the mosaic.

This electron gun consists of an indirectly heated cathode of its own, with the emitting area located in the tip of the gun's cathode sleeve.

The electron stream from this gun is focused by the electrostatic field between the elements of the gun itself and a second anode. The electron gun will be more fully covered when receivers are dealt with.

4. The inner surface of the neck of the iconoscope, as well as part of the sphere, is metallized, and serves as a second anode for the gun, as well as a collector of photoelectrons from the mosaic.

The electron stream from the gun can be deflected by means of magnetic fields at right angles to each other. By means of these sets of coils, two in series for vertical scanning, and two in series for horizontal scanning, each set fed alternating current at a frequency proper to give the "lines" and "pictures" or "frames" desired.

The electron stream plays across the mosaic as a scanning beam. In addition, the alternating current fed the scanning coils is of the sawtooth variety, giving quick return.

The actual working of the iconoscope is best understood by considering the circuit of a single photoelectric element of the mosaic (see Fig. 19).

P represents the element, and *C* that element's capacity to the signal plate, common to all the elements.

The optical image is sharply focused upon the mosaic. Every element *P* on the mosaic emits electrons in proportion to the amount of light falling upon the respective elements.

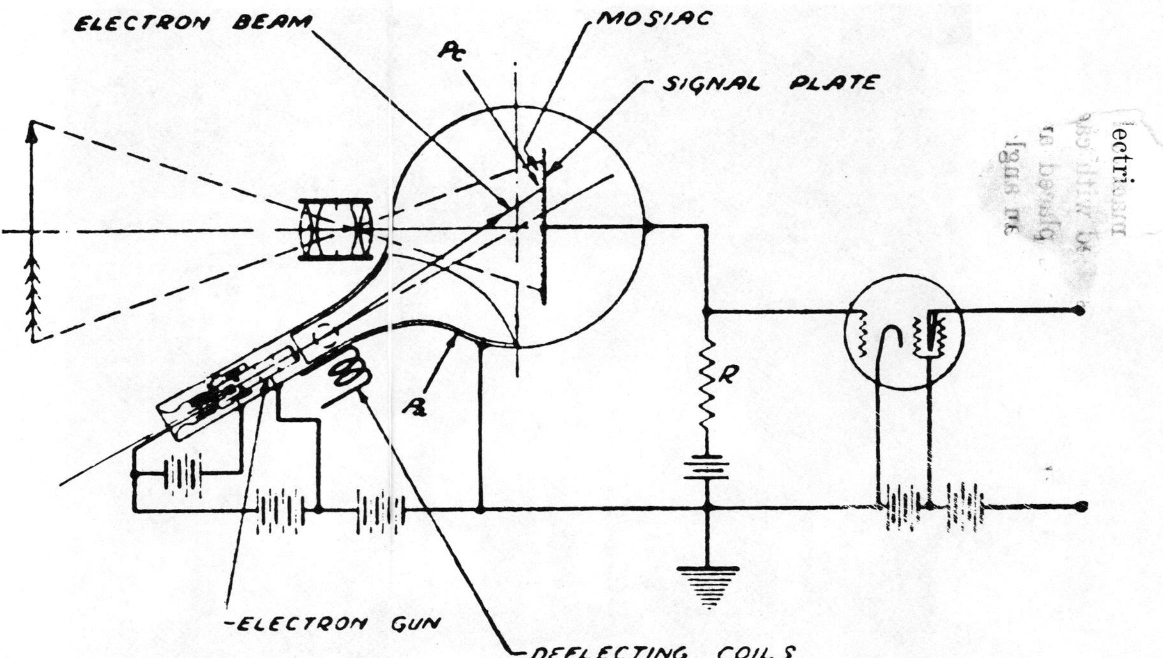

Fig. 19. The Construction of the Iconoscope Showing Circuit of Single Photoelectric Element

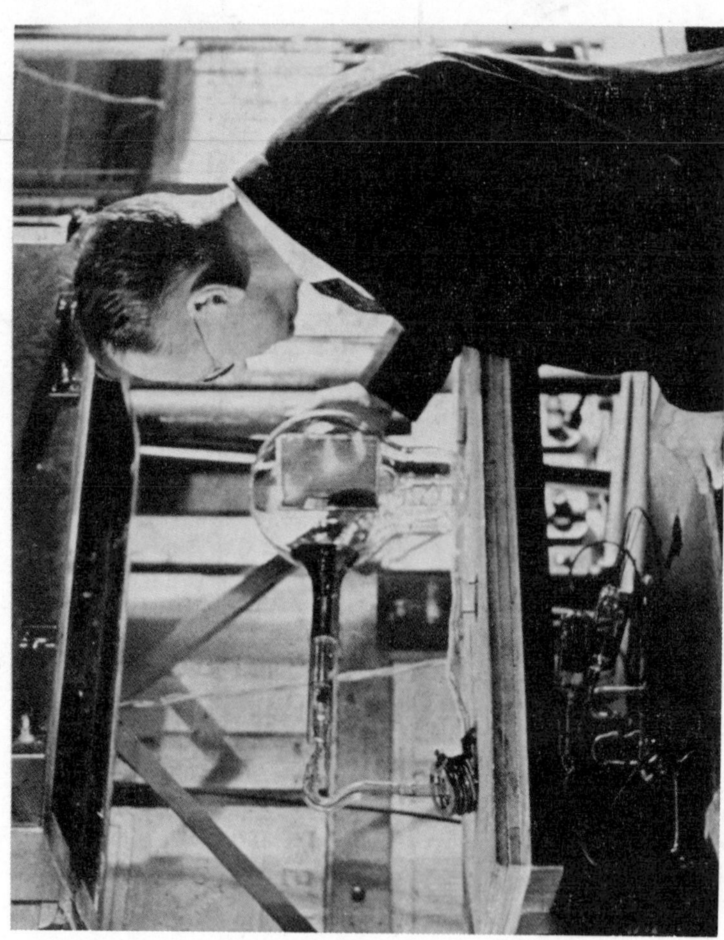

Fig. 20. Dr. Zworykin Shown with the Iconoscope Tube in the Research Laboratory

When the element P emits electrons, due to light falling upon it, electrons (negative) are, of course, given up. This, then, leaves the element charged positive. The magnitude of this charge is also a function of the light intensity, because the electrons emitted were directly proportional to the light.

Now when the **electron** beam which scans the mosaic

Fig. 21. Studying a New Tube

from the electron gun strikes this particular element, the beam supplies the electrons which were emitted, and the element may be said to be discharged.

Electrons from the beam, in excess to those needed to discharge the element, merely "roll" off, as it were, just as a jet of water directed into a partly filled bucket, fills the bucket, and the excess overflows. These overflow electrons are collected by the anode of the iconoscope.

The complete electrical circuit of each element (see Fig.

19) can be traced from P_c (cathode) to C (signal plate), resistance R, the source of an E.M.F., and the anode P_a.

It will be seen that the discharge current from each element will be proportional to the positive charge on that

Fig. 22. The Control Panel of the R.C.A. Experimental Television Station in Camden, New Jersey

element, and hence the intensity of light falling on that element.

The electrical circuit then transforms this discharge current into a voltage signal across the output resistor R.

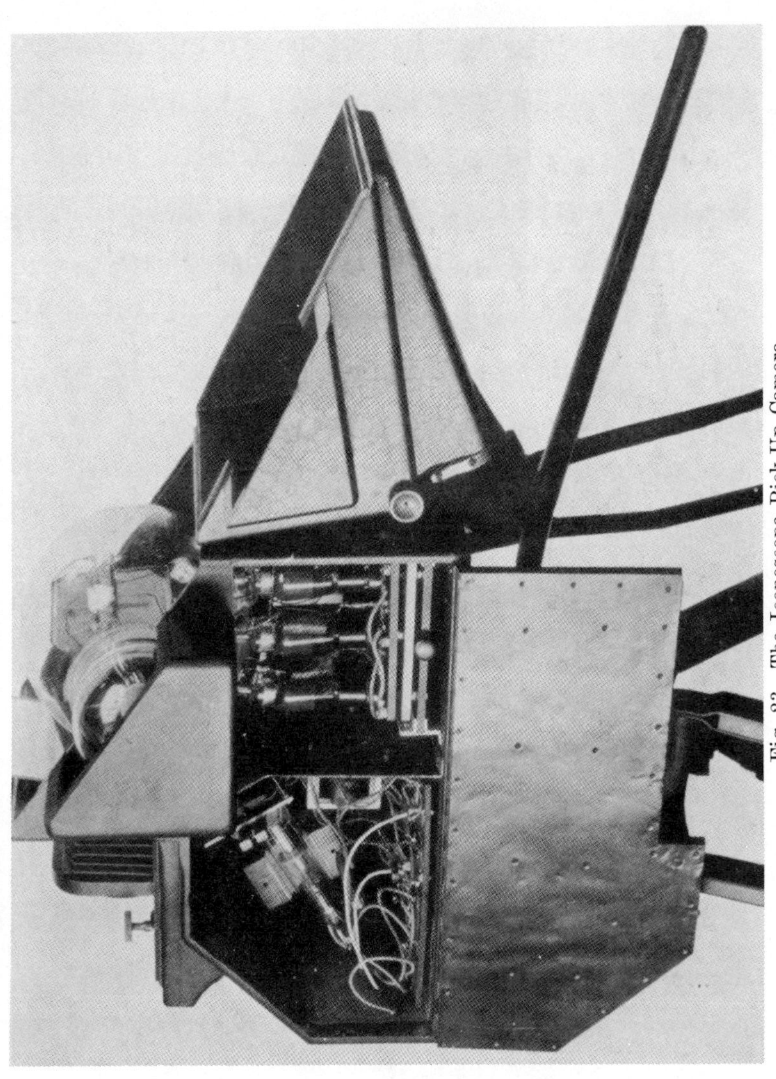

Fig. 23. The Iconoscope Pick-Up Camera

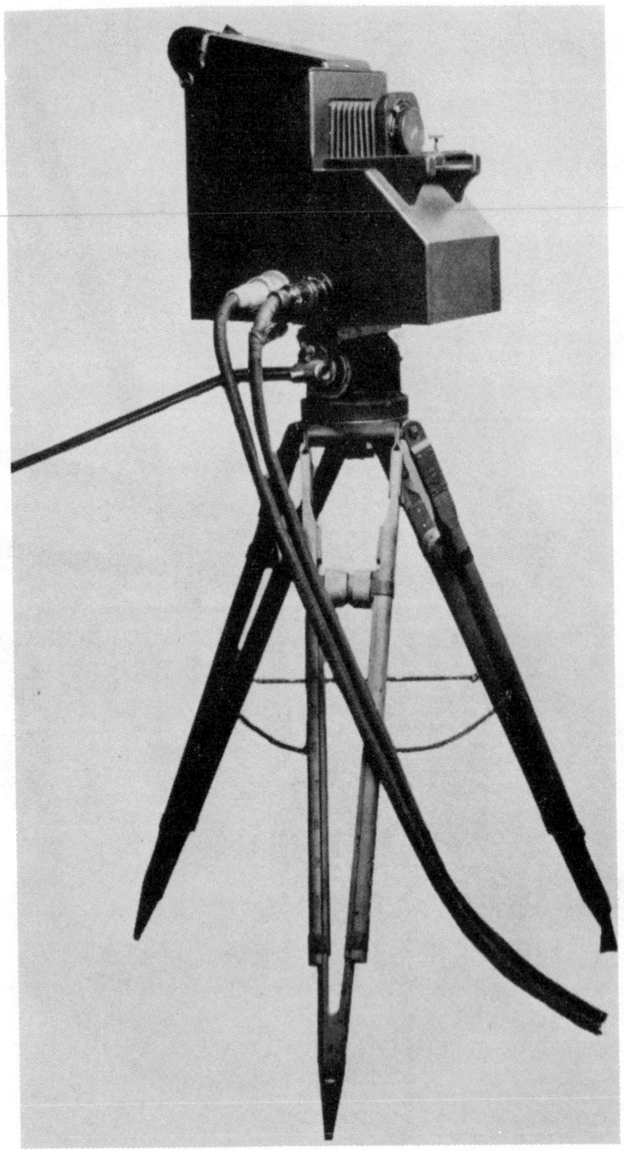

Fig. 24. The Iconoscope Pick-Up Camera Complete

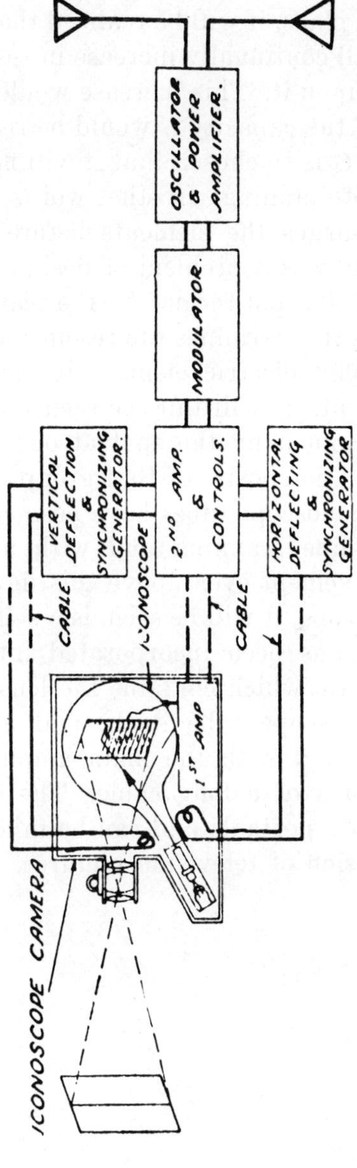

Fig. 25. Schematic Drawing of the Iconoscope

Many difficulties had to be overcome in order to achieve this functioning of the iconoscope.

In the first place, it will be realized that the charge on the element *P* will continually increase in respect to time as the light shines upon it. This increase would continue until the saturation of the capacity *C* would be reached. The saturation point of *C* is so chosen that it will never be reached for a fixed rate of scanning. In other words, the scanning beam always discharges the elements before the capacity *C* is reached. This was a problem of design, which was solved.

The size of the spot formed by the scanning beam is quite important, as it determines the resolution of the iconoscope, since each photoelectric element is much smaller than a picture element. It will thus be seen that a number of elements are scanned by the spot at one time, the resultant signal being a composite of that group.

Actually iconoscope tubes have been constructed to make possible 500-line scanning and with a good margin for future improvement even above this figure.

The iconoscope, it will be seen, is a self-contained pick-up unit; and it has been incorporated into a very compact pick-up camera, which contains the lens, for optical focusing, the iconoscope tube itself, and a pair of amplifier stages, connected with the main amplifier and deflecting units by means of a long cable. The iconoscope pick-up camera is fully portable and can be taken to any point for the transmission of television pictures.

CHAPTER VI

HIGH DEFINITION

Explanation

High definition is a term that will be heard frequently in the discussion of electronic television pictures. The term itself is self-explanatory, since it means a picture at the receiver that is clear and sharp in detail. Before going into the subject of high-definition pictures, it may be well to first study the difficulties involved and then the manner in which electronic television has surmounted them.

Problems

The difficulties involved can best be understood by considering the photographs reproduced in newspapers. Those reproduced in the better newspapers are made up of approximately 10,000 picture elements per square inch. This would be **100 screen,** as the term is used in the printing trade. These individual picture elements can easily be seen in any newspaper picture by using a reading glass. Therefore, if as good a picture is to be obtained in television as is found in the daily newspapers, it will be necessary to build up the reproduced television picture from, and with, 10,000 picture elements in each square inch of picture.

Now to transmit television pictures practically it is necessary that they be transmitted one element at a time; and, as is the case in the pictures printed in the newspaper, the greater the number of elements per square inch for a given picture, that is, the smaller each element, the more excellent is the picture obtained.

One Method of Transmitting a Picture

While the following method of transmitting a picture would obviously not be practical, yet it serves to make

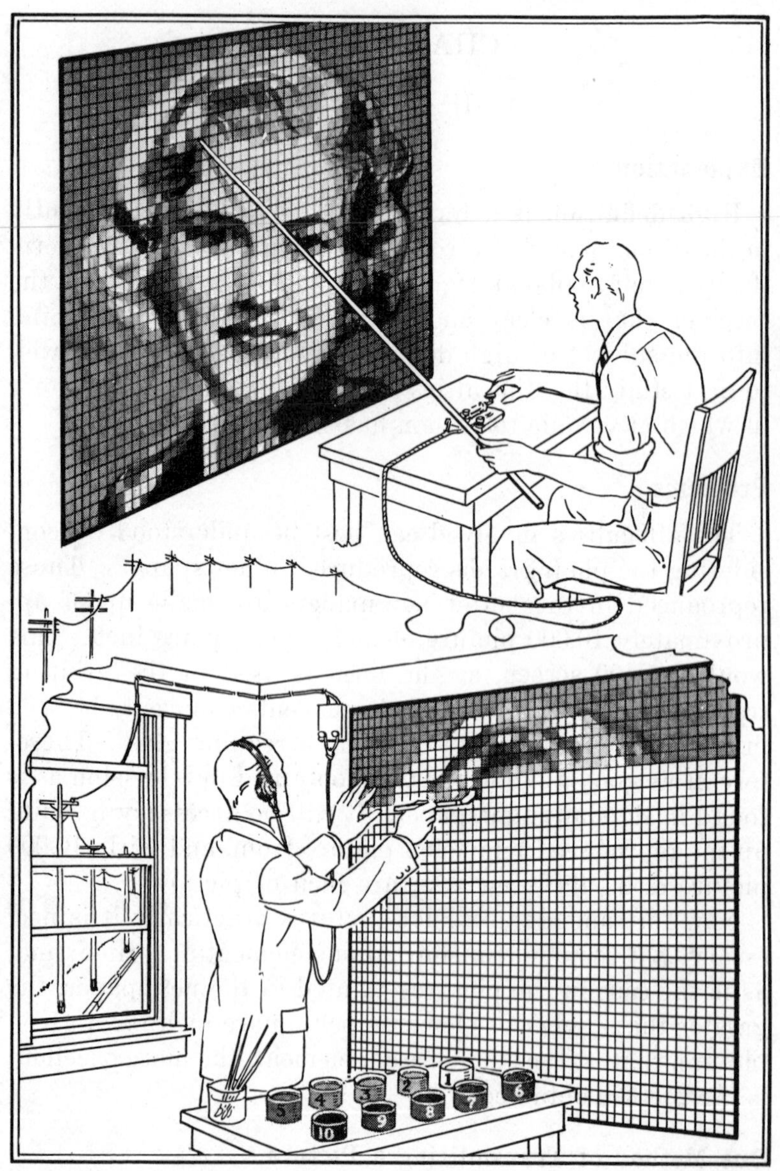

Fig. 26. A Simple Method for Sending Pictures by Wire or Radio

clear a procedure which will help the reader to more easily grasp the superiority of electronic television over other methods.

Taking an average photograph 3 inches square, that is, 9 square inches in area, it will be seen that, in order to transmit this photograph with a definition as clear as a newspaper reproduction, it would be necessary to break it up into nine times 10,000 elements, or 90,000 elements.

It would be possible, but not practical, to do this as follows. The 3-inch-square photograph could be greatly enlarged, and 300 fine lines drawn across it horizontally, and 300 more fine lines drawn across it vertically. This breaks the picture up into 90,000 small squares or elements.

Now each of these 90,000 elements, or squares, of the photograph, is either entirely black, entirely white, or an intermediate shade of black and white. It would be possible to establish approximately ten shades between black and white, and give each shade an identifying number, say numbers between one and ten. Each element could be considered as having an average shade corresponding to one of these numbers.

At the far end of the circuit a draftsman could have a large sheet of drawing paper ruled with 300 horizontal and 300 vertical lines, corresponding to the lines on the the enlarged photograph, and divided into like squares.

A telegraph operator could send a series of 90,000 numbers, each number representing one of the ten shades of black and white for a respective element of the photograph.

Briefly, the operator would go across each of the horizontal lines, from left to right, sending the numbers corresponding to the shades of the squares as he came to them. When he finished completely across the 300 squares of the top line, he would go back to the left, and begin on the 300 squares of the next line, until he had sent 90,000 numbers.

As fast as the draftsman would receive a signal, he would fill in the respective square on his sheet of paper with the shade designated by the signal. At the end he would have a picture which, if reduced to the original size of the 3 by 3 inches, would be a fair representation of the original photograph.

Needless to say, this procedure, while perfectly possible, would be highly impractical, because of the time consumed in sending by telegraph 90,000 shade signals, at the rate of 4 per second.

In television the pictures must be received and recreated at the receiver, at a speed of at least 24 pictures per second, since the illusion of motion is desired, just as it is produced in motion pictures. This is for the same reason that in motion pictures a minimum of 24 pictures per second must be projected, because the power of the eye is such that no flicker is then apparent, and it appears that the picture is on the screen continuously.

Demand for Speed

To transmit at least 24 pictures per second, and reconstruct them at the receiver, that is, one picture every one-twenty-fourth of a second, would mean that the above procedure would necessarily have to be speeded up at least 500,000 times. In short, electronic television was called upon to perform the task one-half million times faster than the "drawing board-telegraph operator" method outlined, if the illusion of motion was to be achieved.

This presented a new problem, and a requirement radically different from any in existing wire line or radio transmission. In the case of the operator sending four shade designations per second by telegraph, approximately 20 electrical impulses per second were required. To transmit by telegraph at the rate of 250 words per minute, a very high-speed circuit, only slightly more than 60 impulses per second are required. Telephone communication requires between 3,000 and 10,000 impulses per second depending upon the excellence of the reproduced speech.

Electronic television presented an entirely new problem. Taking the example given above of 90,000 picture elements for each picture and 24 pictures per second, it would be necessary to transmit over two million impulses per second.

In actual practice, in the latest developments in electronic television, high-definition pictures will probably have 343 lines or more, with 30 picture frames per second. The method by which this is accomplished will be explained later, but this fact will be used, at this time, to make clear the difficult problem involved, and solved.

Actually a square picture like the 3 x 3 photograph referred to is not pleasing to the eye. Experience, both in photography and television, has found that other ratios are more pleasing such as 3 x 4, 4 x 5, 6 x 7, etc. The 3 x 4 ratio will probably become standard in television. This ration of the height to the breadth is known as the **aspect ratio.**

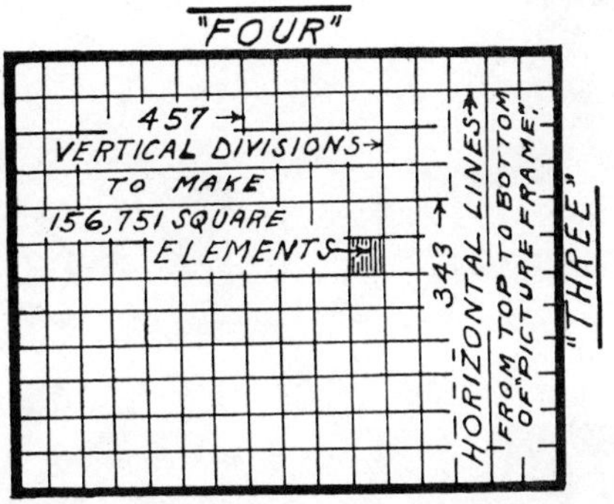

Fig. 27. 343 Lines with Aspect Ratio of 3 x 4

Thus it will be seen (Fig. 27) that if 343 lines are used, 457 divisions would be necessary in order to divide the picture field into square elements. This would make 156,751 elements for each picture, and with 30 pictures per second, the astounding figure of 4,702,530 impulses per second would be required.

The problem confronting electronic television was to speed up the procedure outlined in sending the photograph

Fig. 28. How "Definition" Increases with Number of Scanning Lines

(Courtesy of: R.C.A.)

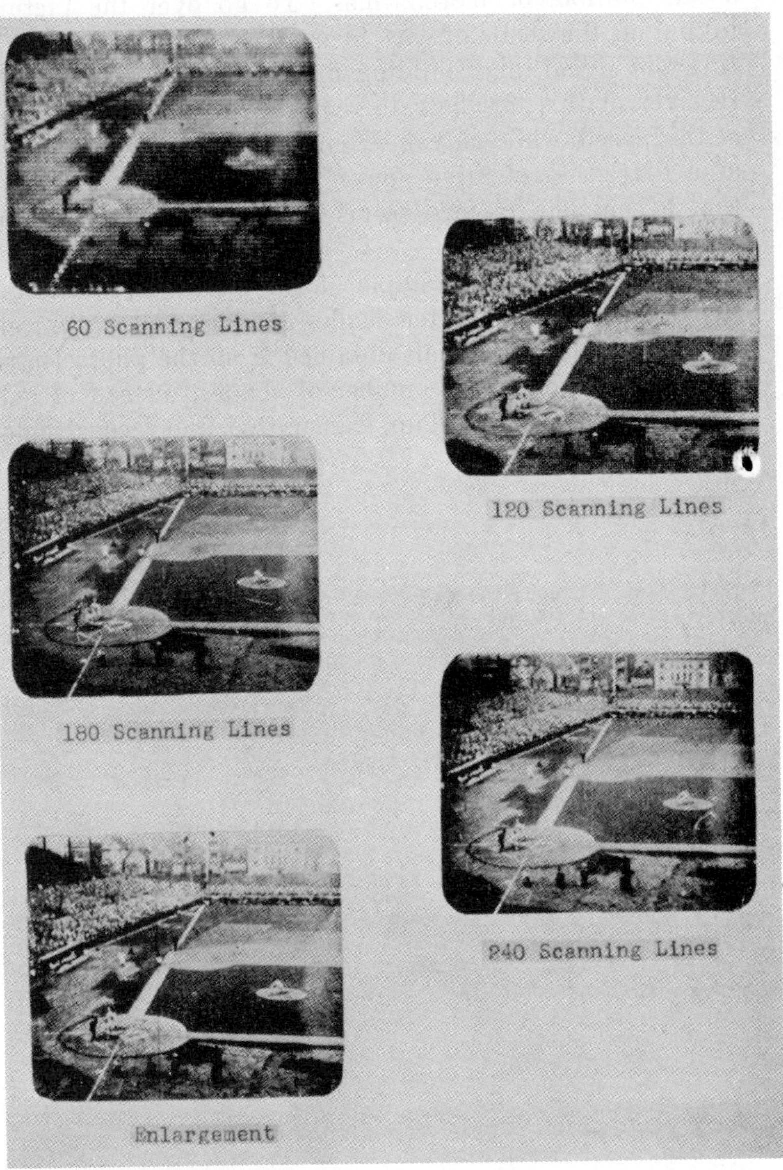

Fig. 29. Television Picture of Actual Outdoor Scene Showing How Lines Increase the Definition

(Courtesy of: R.C.A.)

a half-million, or more, times. To go over the picture, taking off the shade of each element in sequence, transmitting the signal, and building up the picture again in the receiver at this terrific rate required ingenuity. Working at this speed, while only one element is actually reproduced at a time, the retentive power of the human eye is such that a complete picture seems to appear before it at all times.

And again, in the example of the photograph sent one element at a time, only ten shades of black and white were used. The varying signals obtained from the photoelectric surface give an infinite number of shades, instead of only ten, all accurately picked up, transmitted, and reconstructed at the receiver.

CHAPTER VII

HIGH DEFINITION—*Continued*

How the Obstacles Were Met

Experiments showed that at least 200 lines per picture field were necessary in reproducting pictures satisfactory for sustained entertainment. Obviously, the more lines per picture the higher the definition and quality of the picture at the receiver.

As has been stated, 343 lines per picture, 30 pictures per second, interlaced, have been suggested to be used in electronic television. Interlacing will be explained later. First, however, it may be well to call attention to certain features of electronic television which would seem to prove its superiority over older methods involving mechanical scanning.

If 343 lines are used, at a rate of 30 pictures per second, that is, scanning a complete picture every one-thirtieth of a second, each line will necessarily be scanned, or completely gone over from left to right, in less than one ten-thousandth of a second.

The scanning line moves across the picture from left to right, and then very quickly snaps back to the left to again begin its rapid, but uniform, journey across the next adjacent line. In its journey across a single line in $\frac{1}{10{,}290}$ of a second, 457 elements are scanned. This means that each element is scanned in $\frac{1}{4{,}702{,}530}$ of a second. See Fig. 30. These figures present strikingly the great rapidity of scanning. (Actually the scanning time is even less because this includes return.)

In the first place it must be realized that the electron

emissions from the electron image, which are, of course, proportional to the light falling on every respective part of the optical image, are often extremely feeble. The need for an amplifier with quick response at once becomes apparent.

The Electron Multiplier Principle

The quick response of the amplifier used in electronic television, successfully attained by use of the electron multiplier principle, can best be understood by referring back to Fig. 4, and the current wave of a single scanning line. Now suppose that, instead of scanning a picture with varying degrees of light and shade, it were necessary to scan

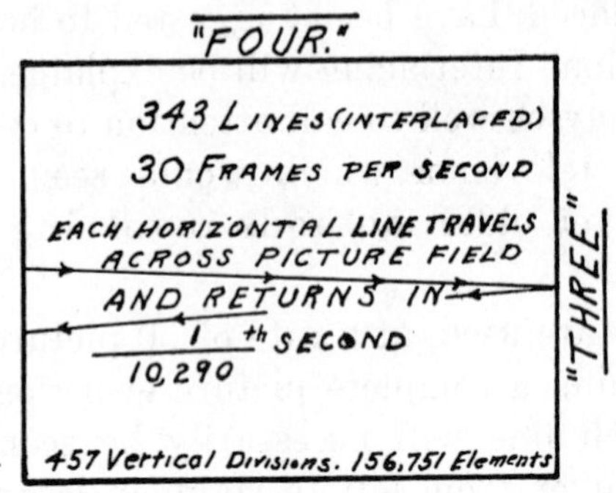

Fig. 30. Showing the Terrific Speed of Scanning

across a white background with a single fine black line upon it (see Fig. 31). From *A* to *B* across the white background there would be a maximum of current in a straight line. Now, upon reaching the fine black line, the current would suddenly drop into a deep valley (*C*), then return to *D*, and proceed along a straight line to *E*. If the line were only one element wide, then it would be necessary for the current to drop to the depth of the valley, and ascend in $\frac{1}{4{,}702{,}530}$ of a second. This time is so short that it is necessary to exaggerate the curve upon the drawing.

If the response were not almost infinitely rapid, it is apparent that one of two things might happen. Either there would be only a very slight valley, as in 1 on the figure. This would only give a very feeble line for transmission and reconstruction at the receiver, and would by no means be high definition. Again, if there was a lag in response, there might be a situation such as shown by 2 on the figure. It is obvious that this would not give a sharp line; it would give a "fuzzy" line at the receiver. But with quick response the line is reproduced sharp and clear.

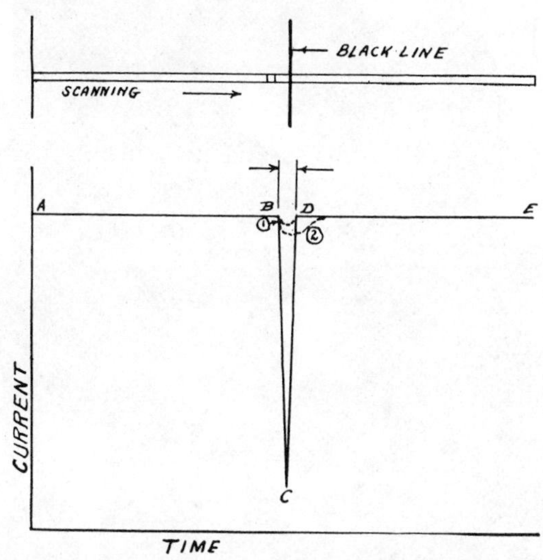

Fig. 31. Scanning across a White Background with a Single Black Line upon It

It will, therefore, be seen that the unusual wave forms typical of television, and the exceeding smallness of the current output from a picture element, combine to make severe demands upon an amplifier if true reproduction is to be obtained.

In electronic television all of the demands outlined have been met. To do the things outlined thus far by means of mechanical scanning would be extremely complicated and cumbersome, if really possible at all. It is obvious that the making of mechanical parts to move at the terrific speeds indicated present problems that would, at least, make the cost of devices using them prohibitive for general use.

Fig. 32. The Interior of a Mechanical Scanning Receiver

(Courtesy of: The Franklin Institute)

Electronic television, however, meets all of these requirements in a manner neither complicated nor costly beyond the reach of general distribution.

Fig. 33. Mechanical Scanning Television Receiver. The large disk is driven by a high-speed motor

(Courtesy of: The Franklin Institute)

CHAPTER VIII

SIMPLE LINEAR AND INTERLACED SCANNING

Many modes of scanning have been tried and used, but two modes seem to stand out above all others, the **simple linear** method and **interlaced.** The first method will be described so that the second can better be understood. Indication are, however, that the interlaced method will be used in electronic television, because it eliminates "flicker" and has other advantages.

Simple Linear Method

Let it be assumed that 240 lines are desired at the rate of 24 pictures or frames per second. This means that in each one-twenty-fourth of a second a complete picture is scanned. In the Farnsworth System this means that 24 times a second the complete electron image is carried across the scanning aperture horizontally 240 times, line by line. In the iconoscope the scanning beam passes across the mosaic 240 times in each one-twenty-fourth of a second. For the purpose of making the proposition more clear let it be assumed that the electron image remains fixed and that the scanning line moves.

Due to the fact that both the horizontal and vertical coils are operating at the same time to move the scanning line, the lines are not exactly horizontal but have a slight "pitch," but all are parallel.

Starting at the upper left-hand corner of the "frame," (see Fig. 34), at the point *A*, the scanning line moves to the point *B* at a uniform rate. At *B* there is a quick retrace to *C*. Then from *C* to *D* the line again moves at a uniform rate. This procedure continues until the point *K* is reached and the picture has been completely scanned: 238 scanning lines and quick returns, making 238 complete cycles (horizontal) have been consumed.

It is obvious that it is necessary to get the scanning line back to *A* to repeat the procedure for the next picture. Thus from *K* there is a quick retrace to a point such as *E*, a scanning line to *F*, a retrace to *G*, a line to *H*, then a retrace to *A*, and all is in readiness to repeat the entire procedure. Two hundred thirty-eight cycles were used to scan

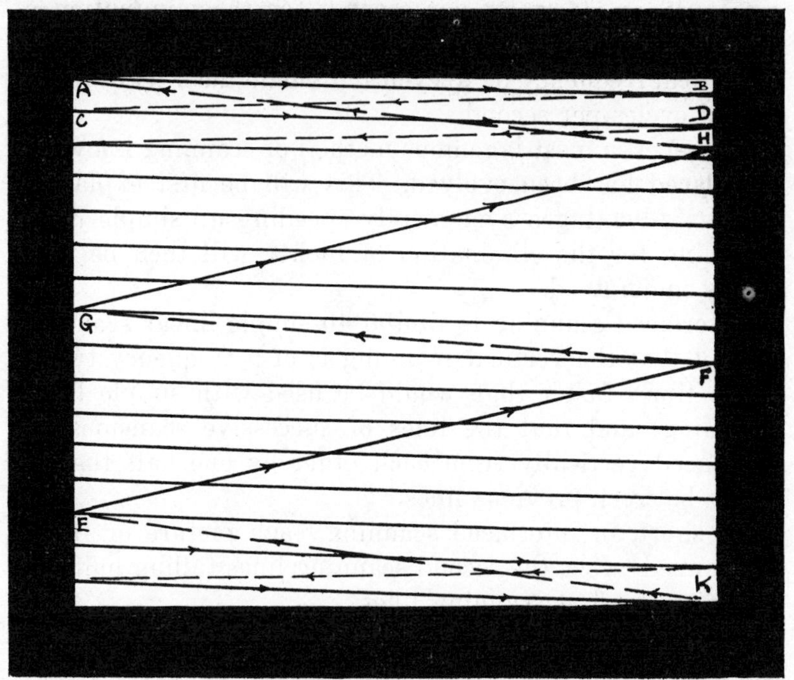

Fig. 34. Simple Linear Scanning

the picture, and 2 cycles to arrive at the point of beginning, thus making 240 cycles or lines.

Interlaced Scanning

It is a well-known fact that if a light source of a given intensity is interrupted at a slow speed it appears to an observer to flicker. However, if the rate of interruption is speeded up, a point can be arrived at where the light appears steady and continuous. The same is true of a television picture, and it has been found experimentally that if the vertical scanning rate is speeded up to the order of 48 to 60 cycles per second, flicker is eliminated.

It is obvious that the vertical scanning rate of 24 cycles per second used in simple linear scanning might be speeded up to 48 or 60 cycles per second, for the elimination of flicker. If the 240 lines per picture frame were used with a vertical frequency of 48 cycles, it would mean 11,520 horizontal cycles per second.

However, a most ingenious method of scanning known as **interlaced** has been evolved. This will be first explained, and its advantages over merely speeding up simple linear scanning for the elimination of flicker will then be more easily understood.

Interlaced scanning is similar to simple linear scanning, except that the vertical scanning is of a frequency two or three times faster than would be used with simple linear scanning; and that the lines of successive scansions are displaced vertically from each other by one half the distance between previous lines.

In short, in interlaced scanning, each picture or frame, is scanned twice, the second scanning lines falling half way between the first scanning lines.

Thus, 343 lines interlaced scanning at the rate of 30 pictures per second, is the same as two half-pictures of 171½ lines each, each half-picture scanned at the rate of 60 pictures per second. The complete scene is scanned once with 171½ lines at the rate of 60 pictures per second, then it is scanned again with 171½ lines at the same rate of 60 pictures per second, the second set of scanning lines falling halfway between the first. Each picture is scanned twice, and each time with 171½ lines in 1/60 of a second, which is equivalent to 343 lines in 1/30 of a second.

All this can best be understood by referring to Fig. 35. From the point *A* the first set of scanning lines starts. The scanning line moves uniformly from *A* to *B*, then a quick retrace to *C,* and so on to the points *E* and *F.* (The first set of scanning lines are in black.) Starting from *F* the line moves from left to right, but instead of completing its course, at *G* it goes to a point *H,* a quick retrace to *I,* and then to a point *J,* midway between the two first scanning lines. It is obvious that *FG* is a "half-line."

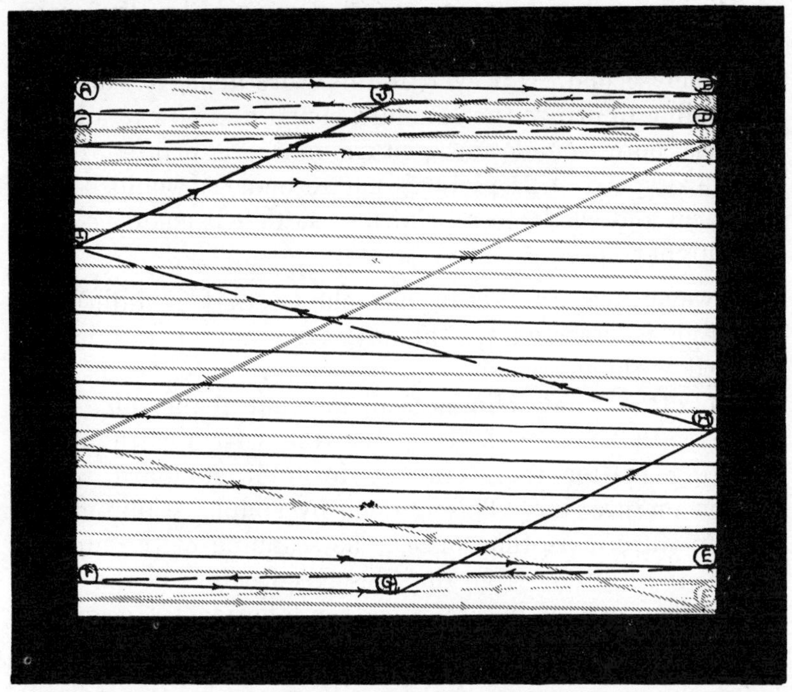

Fig. 35. Interlaced Scanning

Now at *J* the line travels to *B* (gray), and there is the half-line from *J* to *B* (gray). The second set of scanning lines, falling midway between the first are shown in gray. These lines continue until point *E* (gray) is reached. Then the line travels back to *X* (gray), thence *Y* (gray) and finally to *A,* when the whole procedure is repeated.

Now it will be seen that the second set of scanning lines in interlaced scanning fall midway between the first. When the receivers are considered it will be seen that this has an effect like "retouching" or "strengthening" the pictures at the receiver.

There is also another very important reason why it is better to eliminate flicker by means of interlaced scanning rather than speeding up simple linear scanning. It is obvious that in the case of simple linear scanning, 240 lines at 24 picture frames per second, flicker could be eliminated by speeding up the vertical scanning rate to 48 or 60 pictures per second. However, if this were done, it would be impossible to utilize the standard sound film now available, because both sound and picture film are printed for a standard speed of 24 pictures per second. It would be impractical and costly to reprint film so that it could be used at a higher speed.

However, careful study has shown that if interlaced scanning is resorted to, the vertical scanning can be increased without speeding up the picture rate. Thus it is possible to not only eliminate flicker, but at the same time keep available for television use the vast wealth of interesting matter found in the form of standard sound motion picture film. This is done by simply scanning the same picture more than once. The scanning of the film by this method differs from that described before in the Telecine, because now both vertical and horizontal scanning coils are used.

Interlaced scanning can be either of the "half-line" variety described, or "full-line"; in the former case an odd number of lines will be used, and in the latter an even number. The number of lines interlaced and whether they will be an even or odd number is a standard that will eventually be fixed. There is much to be said for both "even" and "odd" numbers of lines; 343 lines at 30 picture frames per second were used here merely as an example. From the following table it will be seen how the definition of the resultant pictures is increased by raising the number of lines. Even figures for lines are used for convenience. (The ratio of the picture is 3 x 4, height to breadth.)

Lines	Elements in each picture frame
120	19,200
240	76,800
343	156,751
400	213,200
500	333,000

CHAPTER IX

THE "SAWTOOTH" WAVE CURRENT

Having considered the rudiments and requirements of scanning, the methods by which the desired results are attained in electronic television will now be taken up.

Discussion

The deflection of the electron beam at both transmitter and receiver is most conveniently obtained electromagnetically. This requires that a "sawtooth wave" current be made to flow through the deflecting coils having considerable inductance.

In Fig. 36, the horizontal scanning will first be considered. The horizontal scanning current is supplied to the two coils shown, which are arranged in series. If an ordinary oscillating current were supplied to these two coils, at the proper frequency, the scanning beam would simply be moved forward and backward, *A* to *B*, and *B* to *A*, in a straight line. The rate would be uniform in each direction.

Now it is highly desirable that the rate from *A* to *B* be uniform, for if the rate of scanning across a picture is not uniform across the entire width of a picture field, the signal will not bear the same relation to light intensity in all parts of the field. This effect is even exaggerated in the reproducer so that the picture appears to be unequally illuminated.

However, it is necessary to get back, on the retrace from *B* to *A*, as quickly as possible, since, as has before been explained, it is desired to approximate a continuous straight line as nearly as possible for all of the consecutive horizontal scanning lines, that is, the lines from left to right. The sawtooth wave current accomplishes this very well.

It is analogous to likening the scanning line, or better, the path of the beam, to a rubber band. A uniformly increasing force is applied to this hypothetical rubber band from *A* to *B*, and the band stretches in direct proportion to the force applied, the greater force being necessary as the point *B* is approached. At *B* the band is released and

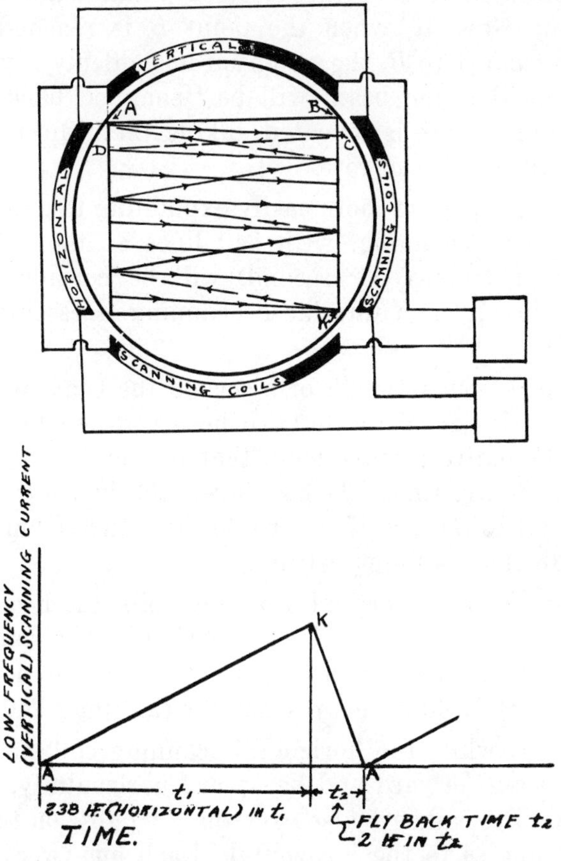

Fig. 36. The Sawtooth Wave Current

it quickly, almost instantly, snaps back to *A*. The increasing force is again applied and the operation repeated.

Now the force of a magnetic field increases in direct proportion to the increase in potential supplied to the coils. Thus, if a uniformly increasing potential were supplied to the horizontal scanning coils, which are in series, the electron beam would be pulled from left to right at a uniform rate as the potential increased uniformly. The two fields being in series, it is obvious that one attracts, while the other repels, so that as long as the potential is in one direction, the motion of the electron beam will be in one direction. Now, if, when the point *B* is reached on the journey from *A* to *B,* the potential is suddenly reversed, it is obvious that the beam will be "snapped back" to *A;* then the procedure is repeated, taking the beam from *A* to *B* again at a uniform rate.

For the purpose of more easily explaining this procedure, simple linear scanning, with 240 lines at the rate of 24 pictures or frames per second, will be considered. The problem involved in interlaced scanning is essentially the same.

Each one-twenty-fourth of a second the beam makes 240 complete trips from the left to the right, and back to the left of the entire picture field, that is, the field is scanned horizontally 240 times. In each cycle the journey from left to right (*A* to *B*) is uniform, while the return from right to left (*B* to *A*) is a quick retrace.

If there were no vertical scanning coils, the beam would merely move forward and backward in a horizontal line, such as *A—B,* completely across the picture field, at a frequency of 5,760 cycles per second (24 times 240).

However, while the horizontal scanning coils are carrying the beam forward and backward horizontally, the vertical scanning coils are also exerting a "pull" on the beam, at right angles to the horizontal. Each one-twenty-fourth of a second the vertical coils carry the beam completely down from the top of the picture field to the bottom, at a uniform rate, since they too are fed a "sawtooth current" of proper frequency, and a quick "retrace" to the top of the picture field.

Thus it will be seen that two "forces," of different frequencies, and at right angles to each other, are "pulling" on the beam. The path of the beam is therefore a resultant of these two forces, and instead of the horizontal lines being exactly horizontal, as from *A* to *B,* they have a slight pitch as from *A* to *C.* Finally the scanning line reaches a point *K*.

In short—always remembering that the purpose is to scan the entire picture field, and to get back to the initial point to begin anew—in simple linear scanning the procedure is as follows (see Fig. 36): For the scanning of one

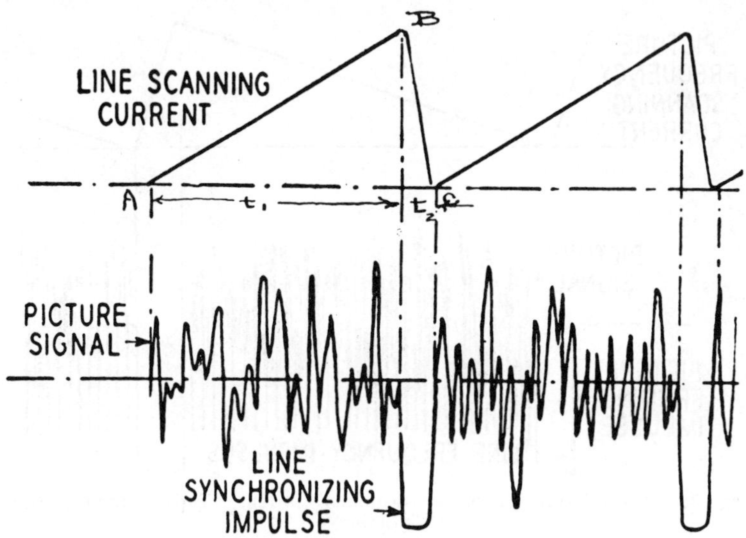

Fig. 37. The Line Synchronizing Impulses

picture frame in one-twenty-fourth of a second, the vertical coils "pull" the beam uniformly from the top to the bottom of the picture field, while the horizontal coils move the beam across the field 238 times. Reaching the point *K,* there is a quick retrace of the vertical scanning from the bottom to the top of the field, and during this retrace, there are two horizontal lines used. This brings the beam back to *A,* the initial point.

It must be remembered that the sawtooth current does not go out with the picture signal. The sawtooth scanning currents, both vertical and horizontal, are generated for the pick-up, and like horizontal and vertical sawtooth scanning currents, identical in respective frequencies, are generated for the receiver in the receiver.

It is obvious, however, that there must be synchronization of deflecting coils between the pick-up and the receiver. This is achieved by means of pulses transmitted between horizontal lines, and between vertical scansions. These pulses are transmitted right along with the picture

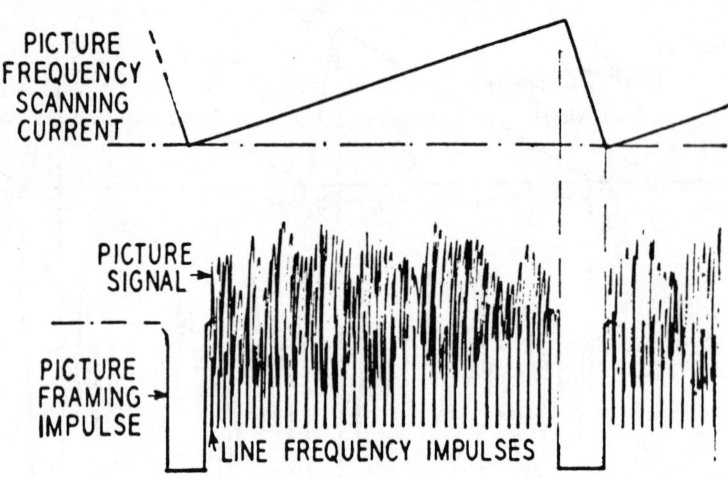

Fig. 38. The Picture Frequency Scanning Current (Top) Complete Signal (Lower)

frequencies, though preferably over a separate radio transmitter. They are separated from the picture frequencies at the receiver simply by amplitude selection.

The period of the quick retrace in both horizontal and vertical scanning lines is most ingeniously used as the period for sending these synchronizing impulses.

In Fig. 37, there will be seen (above) the horizontal sawtooth scanning wave for two complete horizontal cycles. As the sawtooth current rises uniformly from *A* to *B*, in the time t_1, the line moves uniformly from left to right. The picture signal, as shown in the lower part of the diagram is picked-up, and transmitted. This is the signal as shown for one line in the lower part of Fig. 37. At *B* the sawtooth scanning current suddenly reverses, and quickly returns to *C*, in the time t_2. This time t_2 is the period for the quick return, and it is this period that is used to send out the horizontal synchronizing impulse as shown. Thus there are 240 horizontal synchronizing impulses sent out every one-twenty-fourth of a second (in simple linear scanning).

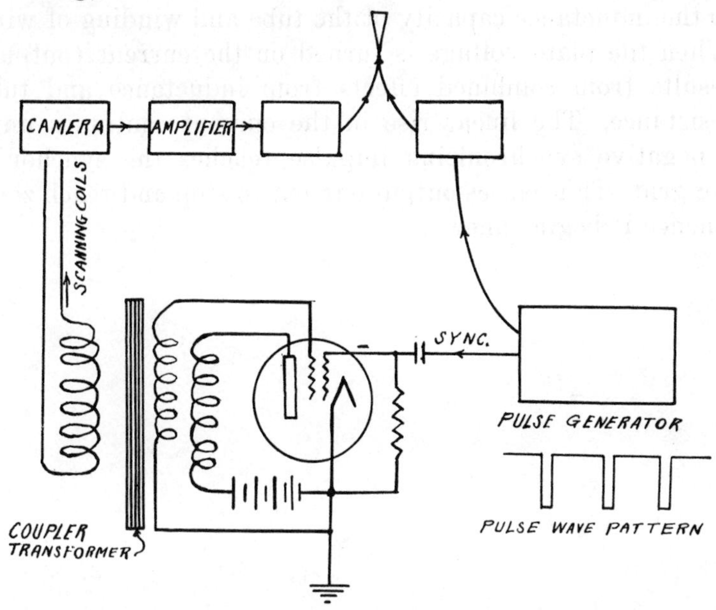

Fig. 39. Circuit of Oscillator Used in Farnsworth System to Supply Sawtooth Current

However, for every 240 horizontal lines, and their respective synchronizing impulses, there is one vertical cycle (see Fig. 38). Between every vertical scansion, and during the vertical "quick return," a "picture framing impulse" is also sent out with the signals, and the lower part of Fig. 38 shows the signal that goes out, with both the horizontal and vertical synchronizing impulses shown, as well as the picture signals.

Many new and interesting tube generators have been developed to generate sawtooth wave currents. The circuit of one of the oscillators used in the Farnsworth System to supply the sawtooth scanning currents is shown in Fig. 39. It employs a special tetrode similar to the ordinary power tubes used in radio receivers except for the grid used for synchronizing.

The circuit is similar to that of an ordinary oscillator circuit, except the capacity of current is small compared to the inductance capacity of the tube and winding of wire. When the plate voltage is turned on the current (output) results from combined effects from inductance and tube resistance. The linear rise in the current continues until a negative synchronizing impulse reaches the synchoniz-ing grid. This causes output current to stop and reach zero, whence it begins anew.

CHAPTER X

THE RADIO FREQUENCIES INVOLVED IN ELECTRONIC TELEVISION

Problem

Now that the optical image has been "dissected" in the pick-up device, and an "electrical replica" obtained in the form of television signals, the next two problems, it will be seen, are (1) How are these signals to be transmitted to points distant from the pick-up? and (2) How are the signals at those points to be received and reassembled into optical images?

The radio frequencies involved in electronic television presented new problems. It must be remembered that in the transmission of speech and music over the standard broadcast band between 550 and 1,500 kilocycles, only 5,000 electrical impulses per second are required, although in some of the present "high fidelity" broadcasting as many as 8,000 impulses are used to give a "truer" and better transmission and reception of music. Taking 10,000 impulses as the outside limit required for the standard broadcasting of speech and music, it will be seen that 10 kilocycles (10,000 cycles) constitute a broadcast band sufficient for each station in the standard broadcast band between 550 and 1,500 kilocycles.

It is a basic principle in radio communication that the width of a frequency band required for a given type of communication depends upon the number of electrical impulses per second which must be transmitted.

The **radio frequencies** involved in high-definition electronic television arise largely from the following factors: (1) **line definition**—the horizontal scanning frequency; (2) **frames or pictures** per second—the vertical scanning frequency; (3) **the aspect ratio**—the ratio of the picture's height to its breadth.

This is better explained by taking a specific case. Let it be assumed that it is desired to transmit 343 lines, interlaced, at 30 pictures per second, with an aspect ratio of 3 (height) to 4 (breadth).

A glance at Fig. 40 will show that it will be necessary to divide the breadth (4) into 457 divisions in order to divide the picture into square elements if 343 horizontal lines are used. The ratio of course is 3 is to 4 as 343 is to 457.

It will, therefore, be seen that each picture frame is divided into 156,751 elements. And for 30 picture frames per second this would mean that to scan the entire picture frame 30 times each second would require 4,702,530 impulses per second to be transmitted.

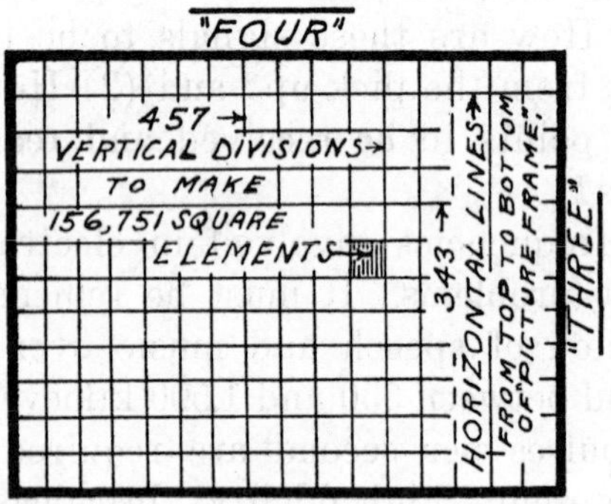

Fig. 40. How Aspect Ratio Governs Number of Elements

Since the entire standard broadcasting band extends only from 550 to 1,500 kilocycles, that is, it consists of only 950,000 cycles, it is obvious that a single high-definition electronic television station would require a band five times as wide as the entire standard broadcast band allotted to all of the standard broadcasting stations.

For this reason it was necessary to utilize wave-lengths in the more unused part of the radio frequency spectrum, that is, in the neighborhood of wave-lengths from ten to one meters.

It may be well to here recall that the frequency multiplied by the wave-length in meters must always equal 300,000,000—the velocity of light and radio waves in meters per second. Or, in other words, if the wave length in meters is divided into 300,000,000 the frequency will be obtained. Thus in the band from ten meters to one meter, there is a frequency range from 30 million to 300 million cycles per second. This gives a wave band of 270 million cycles, and it is apparent that there will be room for the exceedingly wide frequency bands necessary to high-definition television.

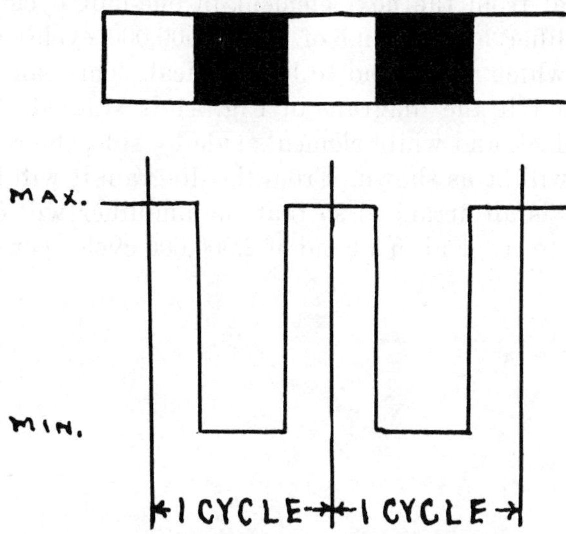

Fig. 41. Amplifier Response

However, it was found that transmission of these very high frequencies differed appreciably from that on longer wave-lengths in that the useful range of transmitting stations was approximately limited by the horizon of the transmitting antenna erected at a high elevation. Recent developments, however, seem to indicate that ultra-high frequency waves carry much farther than was supposed.

It is apparent that the transmission of sound, requiring a band of only 10,000 cycles, would have no crowding effect in the broad band available for television. It is planned to transmit the sound simultaneously with the pictures, but using a separate antenna, and another frequency near that of the picture frequency.

However, the design and construction of amplifiers to respond to a band as broad as 5,000,000 cycles was found to be impractical, to say the least. Therefore, it was so arranged that for every impulse only one-half cycle was necessary, or, in other words, the amplifier response to a change in amplitude of signal arising from one element to the signal from the next element in one-half cycle. Thus the amplifier had a band of only 2,500,000 cycles to take care of, which was found to be practical. This can best be understood if the diagram in Fig. 41 is studied. Taking totally black and white elements side by side, the resulting current will be as shown. From the diagram it will be seen how this is all arranged so that the amplifier will only be required to respond to a band of 2,500,000 cycles per second.

CHAPTER XI

AERIAL EQUIPMENT FOR USE IN ELECTRONIC TELEVISION

Since high-definition electronic television will be necessarily transmitted in the ultra-high frequency range of the radio spectrum, the bands from 10 meters to 1 meter, the subject of antennas for both transmission and reception of electronic television signals has much in common with the study of antennae used in other ultra-frequency work.

The sound part of the program will be transmitted over a frequency band close to that of the vision part of the program, and while separate antennas will be used at the transmitting station for sound and vision, at the receiver one antenna will be sufficient for both the sound and pictures.

The Dipole Aerial

The great interest, of course, will be in aerial equipment for reception. This will not be at all complicated, and will consist of a dipole aerial and shielded lead-ins. Essentially the dipole aerial consists of a stiff wire one-half, or one-quarter, the length of the wave to be received. Co-axial cable for lead-ins can even now be purchased at reasonable prices and in convenient lengths. This co-axial cable not only shields the incoming electronic television signals from outside interference, but also prevents the signals themselves interfering with other radio devices nearby.

The great source of anticipated interference is from ignition systems of automobiles. This can be largely overcome by placing the aerial well back from thickly traveled roads.

The behavior of ultra-high frequency waves is still a subject of deep study, and undoubtedly many revealing discoveries will be made as this field becomes of more and more importance through television.

A peculiarity of reception of ultra-high frequencies is that "dead spots" are frequently found. Since these are usually quite small in area they can be avoided by placing equipment in a different part of even a small room.

The proper placing of the aerial for electronic television reception will probably be a matter that will require thought and experimenting in each individual case in order to obtain the best results, but the problem is never one difficult beyond solution.

The antenna used for reception of high-definition electronic television does not differ materially from antenna already developed for the reception of ultra-high frequency signals. This equipment is already on the market for ultra-high frequency radio reception.

Fig. 42 shows such an antenna that has given satisfaction and which is quite reasonable in cost. The wooden frame is built as shown. The insulator (*A*) is standard. One-quarter inch copper rod or tubing is used for the dipole. The unit shown at *A* is a bee-hive insulator with receptacles, fitted with thumb screws, to receive the lengths of copper rod. Inside this insulator the two ends of the copper rod are connected by coils of wire as shown. Then a coil of a few turns of wire about this first coil has two terminals with binding posts on the side of the insulator. By this ingenious device a coupler is formed between the dipole and the leads. This device *A* has bracket connections for attaching it to the frame.

The whole length of the rod must be equal to one-half the actual wave-length to be received. By the use of *A* this is easily accomplished.

The lead-in wires are crossed as shown in the sketch. This crossing of the wires is accomplished by using a special spacing insulator which is also standard equipment, and which can be reasonably purchased. This crossing of the

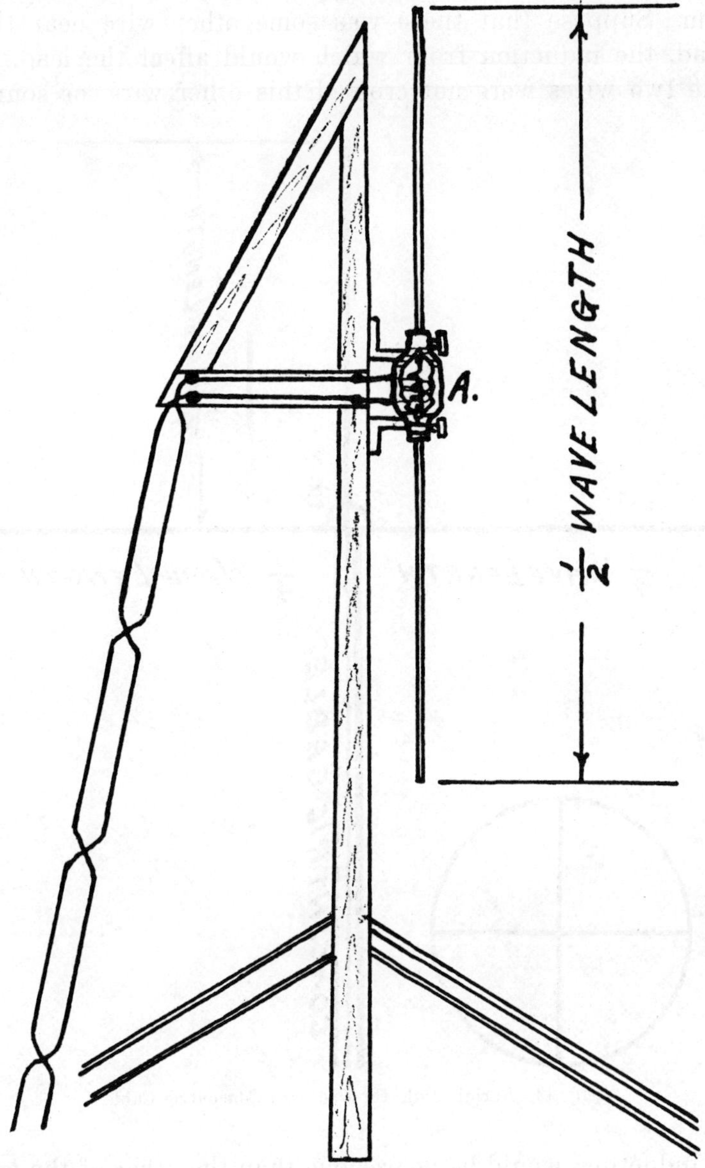

Fig. 42. Di-Pole Antenna for Electronic Television Reception

wires is quite an ingenious way of getting around a problem. Suppose that there was some other wire near this lead, the induction from which would affect the lead. If the two wires were not crossed this other wire, or source

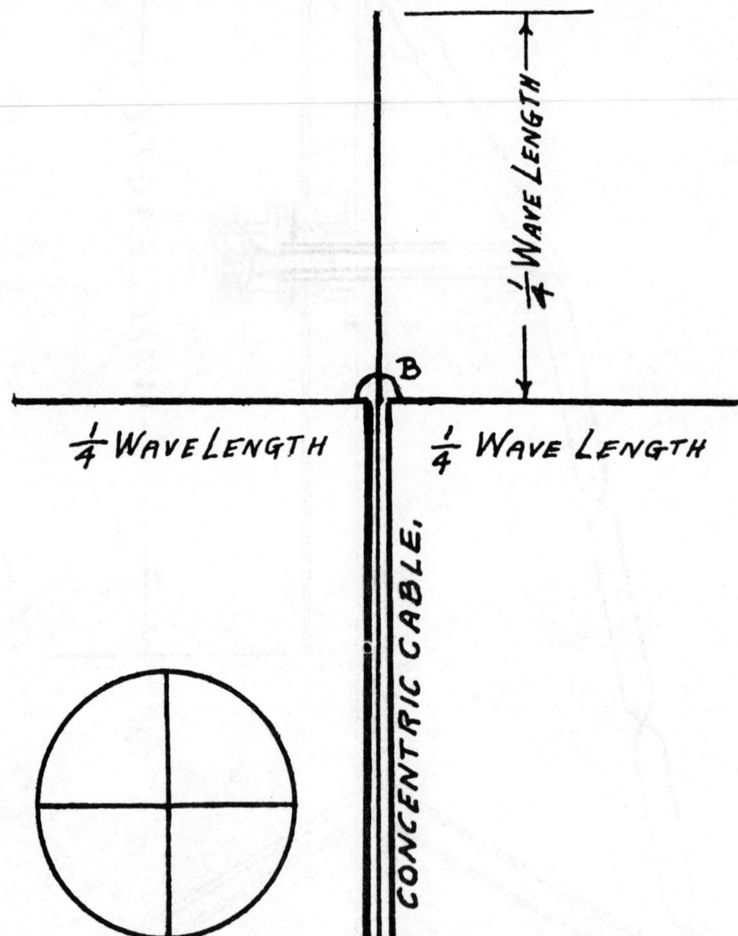

Fig. 43. Aerial Made for Use with Concentric Cable

of induction, would be nearer one than the other of the two lead wires. But with the lead wires crossed as shown this is compensated for. This aerial shown in Fig. 42 can be made at the cost of a few dollars from material already on the market.

Another type of aerial that gives very good results, and which uses concentric cable for its lead is shown in Fig. 43. Here the copper rods are each one-quarter an actual wave length. One-quarter the wave-length is approximately three feet six inches. One rod is placed vertically as shown and connected to the inner wire of the concentric cable. Then four other rods are placed vertically like the spokes of a wheel and these four are connected to the outer "pipe" of the concentric cable. For convenience, and to keep them spaced, the horizontal rods have a rim placed around them as shown in the small sketch. *B* is a special insulator which makes a watertight joint at the end of the concentric cable. Both of these aerials are easily made.

CHAPTER XII

THE TRANSMISSION OF ELECTRONIC TELEVISION OVER LAND WIRES

The Problems Involved

In chapter X it was shown that extremely wide frequency bands are required for the transmission of high-definition electronic television signals—bands of the magnitude of two and one-half million cycles. Therefore it was necessary to look toward the ultra-high frequency bands of the radio spectrum—the bands between thirty million and three hundred million cycles—for radio transmission. Transmission of frequency bands of this magnitude over existing land wires was impossible.

Existing carrier systems were, of course, primarily designed for the transmission of telephone conversations. However, when it became necessary or expedient to transmit broadcasting programs over land wires in order to effect chain hook-ups, the problems encountered were comparatively simple. The ordinary telephone speech circuit requires a frequency range of from 250 to 2,750 cycles. The ordinary wire line for broadcasting requires a frequency range of from 50 to 5,000 cycles, and the highest class wire line for broadcasting requires a frequency range of from 50 to 16,000 cycles.

The present standard open-wire carrier systems transmit three telephone conversations in each direction on each pair of wires, making a total of six channels occupying a frequency range up to 30,000 cycles. By placing the conductors in the lead sheath of a cable, systems have been set up which yield as high as twelve channels on each pair of conductors, and which require frequencies up to 60,000 cycles. From this general outline of the status of the telephone carrier circuits, it will be seen that they are adaptable to the transmission of broadcast programs.

Basically, most probably, research studies on carrier systems had for their chief objective the making available of more channels on each pair of wires, for the wider the frequency band, the greater the number of telephone conversations each pair of wires could carry. If a carrier system could be perfected which would make available a frequency band of the magnitude of one or two million cycles, that system could accommodate a great number of telephone conversations. In addition to this, a carrier system which could accommodate a frequency band of this width could also transmit electronic television of high definition. These circumstances spurred on research in this field, and the answer seems to have been found in the **co-axial cable** (see Fig. 44).

The Co-axial Cable

Returning to the telephone problem, it was found that when more than six channels were imposed upon the present open-wire carrier systems, and the upper frequency correspondingly extended, interference and cross-talk became a serious problem. Then it was found that this interference might be greatly reduced by placing the conductors within the lead sheath of a cable. While interference from outside of the sheath was thus reduced, cross-talk between the circuits within the sheath became a greater and greater problem as the channel frequency was raised.

It might be said that the transmission of electronic television over wires was a secondary thought in the elaborate research that has been pursued on the co-axial cable—a research that has emerged from the laboratory, since an experimental co-axial cable is to be built between Philadelphia and New York by the American Telephone and Telegraph Co.

This would seem economically sound since electronic television will thus be able to utilize special co-axial cables that have for their primary purpose telephone transmission of many channels. In short, the special cables necessary for the wire transmission of electronic television can also be used for telephone transmission, although of course not at the same time.

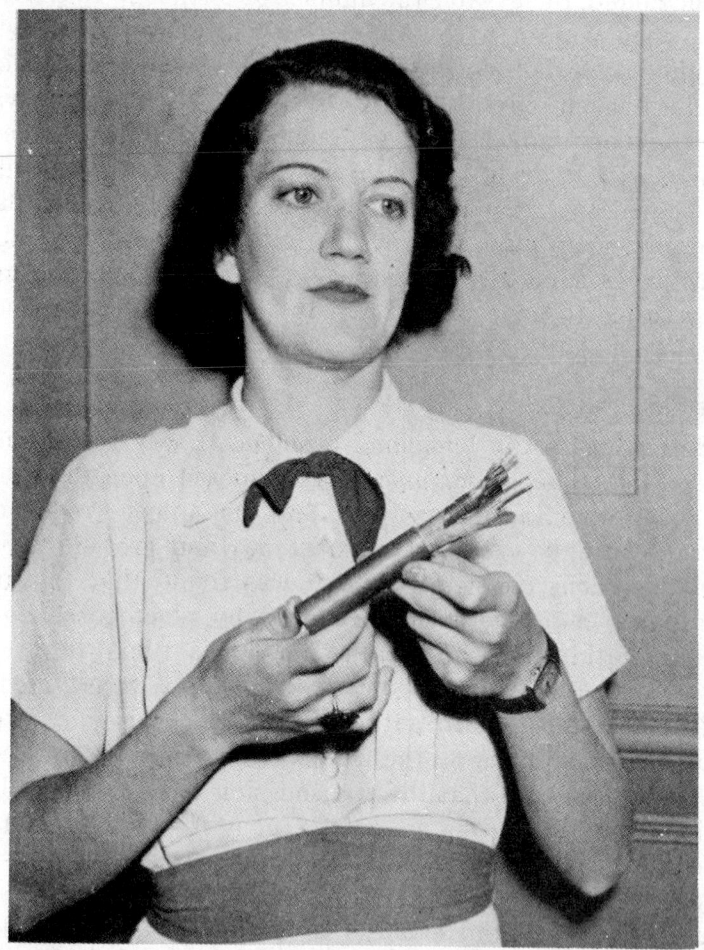

Fig. 44. Piece of Co-axial Cable. The two large tubes can be used for television circuits and the eight other wires are for signaling and control circuits

(Courtesy of: The Franklin Institute)

To return to the engineering rather than the economic side of the problem, research engineers found promise of solution in a single small carrier well shielded. A carrier of this type would be suitable for frequencies as high as desired, because the protection of the shield against interference becomes greater as the frequency increases. Also, a single repeater could be used for the entire frequency band making up a group of channels, instead of using a repeater for each pair of conductors.

The answer to all of this was found in placing a wire within a shield, using the shield itself as one of the conductors. This is the co-axial conductor.

Fig. 45. The Co-axial Cable. The inner conductor has been drawn out of its shield to show how the small insulators are evenly spaced. This shield is made out of flexible aluminum tubing. Dr. C. D. Haigis of Haigis Laboratories, Inc., is demonstrating

Even with the co-axial cable, however, repeaters will be required at intervals of about ten miles. This would make the cost of construction rather prohibitive were it not for the fact that a single repeater amplifies the entire band, and repeaters have been designed which are fully automatic in operation. Experimental repeaters have been designed and built for frequencies as high as 5,000,000 cycles.

The co-axial structure has also been employed for antenna lead-ins, and for this purpose it has been designed to have largely air insulation to minimize high-frequency losses. The shielding effect is highly important because the co-axial design shields the lead-in from outside interference, and also prevents it from exerting interference outside. Co-axial cable of this type will find wide use as antenna lead-ins in electronic television.

PART II

THE RECEPTION OF ELECTRONIC TELEVISION PICTURES

CHAPTER I

THE ELECTRONIC TELEVISION RECEIVER

The "heart" of the **receiving system** is a special cathode ray tube, a vacuum tube having an **electron gun** and a **fluorescent screen.** There are several points of difference between these tubes and the ordinary cathode ray oscilloscopes; for instance, the cathode ray tubes used in electronic television receivers have an added element to control the intensity of the beam.

The general design of the electronic television receiver can probably be best understood by first explaining the various parts and their respective functions. Then the various systems can be taken up in detail.

The **cathode ray receiving tube** is a high vacuum tube as shown in Fig. 46, which is furnished with an electron gun. This electron gun is made up as follows:

1. **The cathode** itself, which is coated with the usual barium and strontium oxides. This cathode may be heated with either D.C. or A.C. current. When the cathode is heated it emits electrons, and a "mist" of electrons is created about the cathode.

2. **The anode.** It is necessary to produce a "stream" from this electron "mist," and give that stream velocity. For this purpose an anode or accelerator is placed before the cathode. In its simplest form this is merely a disk with a small hole in the center. The shape of the anode is important and will be discussed later. A second anode may be formed by a metallic coating in the tube, which will also be discussed later.

3. **The grid.** When the cathode is heated by a constant current, and a steady and unvarying stream of electrons come off, the intensity of the electron stream is always the

same, as in the oscilloscope. However, in the cathode ray tube used in the television receiver, it is desired that the electron stream vary with the incoming picture signals so that the resultant effects on the fluorescent screen will also vary with the respective signals. To achieve this a grid is placed between the cathode and anode. This controlling element corresponds to the grid in the ordinary triode.

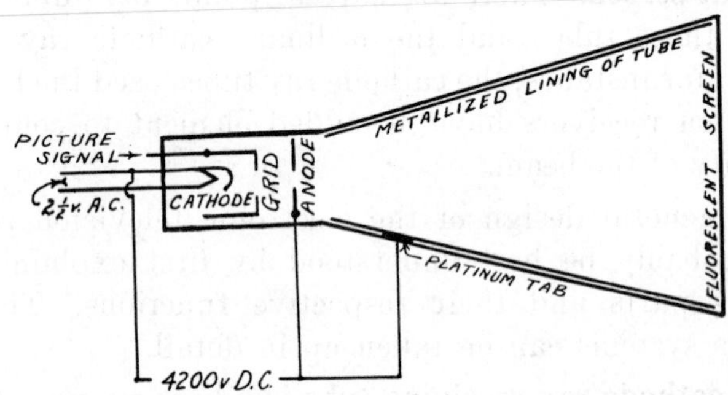

Fig. 46. Schematic Diagram of Typical Cathode Receiving Tube

The electron stream may be likened to a stream of water issuing from the nozzle of a hose. The intensity of the stream will not vary, but if the hose itself is squeezed with varying pressures, the intensity of the stream of water will vary with the amount of "squeeze." The grid acts in an analogous manner, and the electron stream varies with the potential of the incoming picture signal, and in this way the electron stream is modulated.

The fluorescent screen: This is situated on the inner surface of the end of the tube. It is obviously necessary to reproduce light from the television signals received in the reproducer. Three methods suggested themselves: (1) incandescence, (2) ionization of gases, and (3) fluorescence. Incandescence was at first impractical because of the tremendous power needed to heat a screen large enough for television. Ionization of gases was too sluggish to operate well at high scanning speeds used in electronic television, and the beam could not be controlled. Fluorescence has the great advantage of responding to an excitation of very short duration. The fluorescent screen has the property of absorbing electrical energy and emitting light. The screen is very thin so that a large portion of the emitted light passes outside of the tube.

There are a large number of fluorescent materials. The most favored thus far has been synthetic zinc silicate closely resembling the composition of the mineral willemite. Two qualities effected this choice: (1) it is highly resistant to chemical decomposition under electron bombardment, (2) the bright green spot which it reproduces is near the color range to which the human eye responds most strongly. Again, and this is most important, the intensity of illumination is proportional to the current of the cathode ray.

A strong cathode ray will produce a fluorescent light intensity of several candle power, which is sufficient in a dimly lighted room. However, other fluorescent materials have been experimented with, and "black and white" pictures have been obtained.

It is desirable that the average brilliancy of the picture be adjustable. At first it would seem that this could be accomplished by regulating the current heating the cathode, for the greater the heating current, it would appear, the hotter the cathode and the more the electrons emit. However, the problem is not so easily solved, because after reaching a certain point the increase in emissions of electrons is small compared to the increase in current. The problem is solved by regulating the "grid bias" (potential on grid). Going back to the modulation of the cathode ray by the grid being likened to a hose with varying "squeezes" exerted upon it to vary the intensity of the stream, another point of "squeeze" is simply added which controls the total intensity of the stream which is being modulated. In short, the intensity of the stream being modulated is controlled.

It is also necessary that the cross-section of the cathode

ray beam be as small and sharp as possible, in fact, it should be as small in area as the area of a single picture element in the pick-up, in order to obtain a definition in the reproducer as high as that in the pick-up.

To obtain this result it is necessary to focus the beam. This can be done either *electromagnetically* or *electrostatically*. Both methods will be described later.

With the cathode ray beam now sharply focused in the center of the fluorescent screen, and the intensity of the resulting illumination varying with the incoming picture signal, all that remains to achieve a television picture is to move the spot across the fluorescent screen in a manner identical with the scanning motion (both horizontal and vertical) in the pick-up.

The cathode ray beam in the receiver can be deflected by coils just as the scanning beam is deflected in the iconoscope, or the electron image is deflected in the Farnsworth dissector.

Again it becomes necessary to produce sawtooth wave scanning currents in the receiver, so that the spot resulting from the cathode ray beam striking the fluorescent screen may be deflected both horizontally and vertically at the same respective frequencies as those used in the pick-up.

The sawtooth scanning currents in the reproducer can be achieved either electromagnetically or electrostatically.

The synchronizing impulses serve a double purpose in the receiver: (1) they synchronize both the respective horizontal and vertical scanning between the pick-up and the receiver; and (2) they extinguish the spot in its quick retraces in scanning. This latter feature is a great advantage toward procuring good pictures in the reproducer.

CHAPTER II

THE ELECTRONIC TELEVISION RECEIVER —*Continued*

Schematically the electronic television receiver is essentially as shown in Fig. 47. The picture signal enters the radio receiver and amplifier. This signal is made up not only of the picture signals themselves, it will be remembered, but also of the horizontal and vertical scanning impulses.

Filters permit the horizontal and vertical scanning impulses to be separated by amplitude selection from the picture signal proper, and to pass to their respective scanning generators. In these generators, which generate the saw-tooth wave currents needed for the horizontal and vertical scanning coils, these impulses serve to synchronize the respective scanning currents, so that they, in turn, are "in step" with the scanning currents in the pick-up.

The entire signal, made up of the picture signal proper, as well as the vertical and horizontal impulses, passes on to the grid of the electron gun in the cathode ray tube, where it serves to modulate the cathode stream, which in turn impinges upon the fluorescent screen, giving the picture.

The horizontal and vertical impulses in the signal, acting upon the grid, serve to extinguish the cathode ray beam during the retraces in scanning, thus eliminating retrace lines streaked through the picture.

In the outline of an electronic television system shown in Fig. 47, neither the facilities for focusing the cathode ray beam, nor the anodes for accelerating it, are shown. It must be remembered, however, that both of these are highly important, and they will be discussed in their proper place.

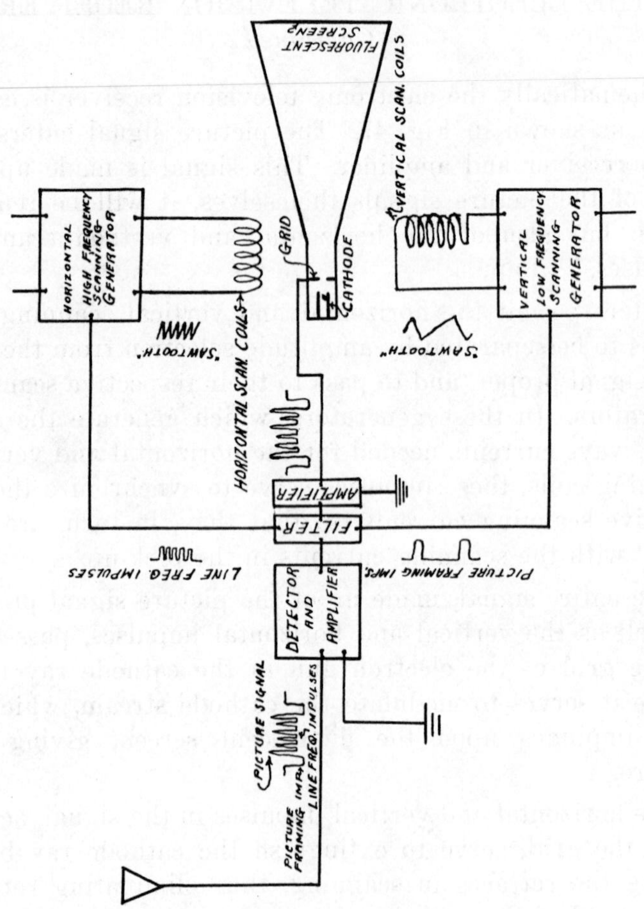

Fig. 47. Schematic Diagram of Receiver

The question naturally arises, How many controls will eventually be necessary in the construction of an electronic television receiver? In short, How many dials and knobs will the viewer be called upon to adjust? Obviously the purpose will be to make the tuning and control as simple as possible.

In the first place it will be remembered that the "sound" part of an electronic television receiver will be separate and apart from the television, although mounted in the same cabinet. Sound will be transmitted and received on a frequency band next to that of the television, all in the ultra high frequency band.

Certain standards will be fixed, the principal ones being the horizontal and vertical scanning frequencies, the "aspect ratio" of the picture field, and the method of interlacing. Once these standards have been fixed sets will be built to conform to them.

With these standards set only three controls will probably be necessary for a sound-television receiver:

1. **Frequency tuning dial.** Since the sound band as well as the television band for a given television station will be fixed, it is apparent that it would not be difficult to have one dial tune in both the television and sound. This involves merely a problem in design.

2. **Sound volume control.** This will be the same as the volume control on the present sound receivers. In fact the sound part of a television-sound receiver will be a separate unit inclosed in the same cabinet.

3. **Control for intensity of the image.** Some people will prefer a "brighter" picture than others, and there will be a control to take care of this. In short, the control of intensity of the picture is quite analogous to sound-volume control—it will vary with the taste and desires of the viewer. The varying degrees of intensity will be achieved by varying the amplifier gain governing the amplification on the incoming signal.

There will be two auxiliary adjustments which have to do with quantities and, when once adjusted properly by service men, need not be varied. These will be:

1. **Adjustment of the "grid bias."** This will control the background intensity of the picture, while the control knob on the front of the cabinet will control the variations of this "background" intensity desired by the viewer. It is very much like a master cock from a water main controlling the total amount of water permitted to flow into a house, while a spigot would control the varying amounts of this total that a consumer would desire.

2. **Adjusting the focus of the cathode ray beam.** While it is highly essential that the cathode ray beam be accurrately focused, for the reasons stated, yet once focused, it should, for all practical purposes, maintain this focus.

CHAPTER III

THE FARNSWORTH RECEIVING SYSTEM—THE OSCILLIGHT

The vacuum tube in the Farnsworth System is known as the **oscillight**, and the system comprises this vacuum tube having an electron gun and a fluorescent screen, positioned in a focusing and deflecting coil system (Fig. 48).

The electron beam from the gun is focused into a small spot on the fluorescent screen electromagnetically by means of a short coil around the neck of the oscillight tube, this coil extending one-third to one-half the total length of the tube.

The horizontal deflection of the spot for scanning is accomplished by small coils made to conform closely to the neck of the tube, inside the focusing coil. Vertical deflection is accomplished by means of the electromagnet shown.

The electron gun in the oscillight (Fig. 49) consists of a heater-type cathode having a concave surface that faces the fluorescent screen. A grid is placed over this cathode, consisting of a small cylinder closed at one and having an aperture in the closed end through which the electrostatic field of the anode may pass to the cathode.

The anode is of the truncated type. The purpose of the "protuberant" end on the anode is to stimulate a point of charge from which the lines of force originate. This makes it unnecessary to carefully align anode and cathode.

The design is such that practically no electrons strike the anode at all, in spite of the fact that the aperture in the anode is only .020 inch in diameter, and that the maximum beam current is more than 15 milli-amps. With a suitable magnetic focusing system the size of the spot obtained from this electron gun is less than .010 inch in diameter.

Now it is obvious that with the electron stream bombarding the fluorescent screen, a charge would build up in the screen end of the tube. The problem that presented itself was how this charge could best be removed, for it is apparent that it must be removed. The electron stream from the gun "splashed out" secondary electron emissions from the screen, therefore some means had to be provided to draw these secondary emissions from the screen.

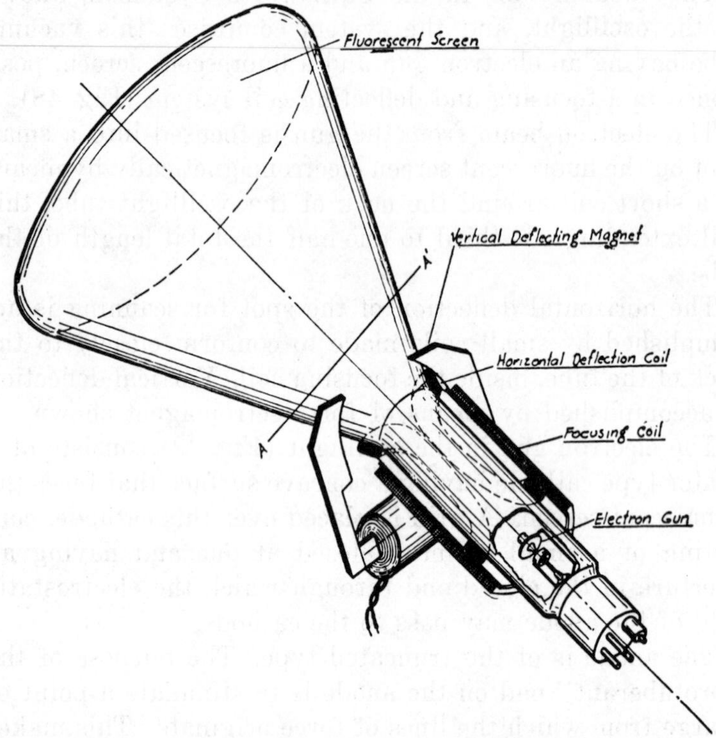

Fig. 48. The Oscillight of the Farnsworth System Shown with Focusing and Deflecting Coils

In the oscillight this is accomplished by evaporating a thin film of nickel on the inner walls of the tube, and to connect this metal coating with the anode.

Magnetic Focusing of the Beam

For the best magnetic focus the solenoid should extend the full length of the electrons' path; but this gives poor deflection sensitivity. It has been found experimentally that the use of a short focusing coil gives entirely adequate focusing of the spot, and that the spot can be made even smaller than necessary. The sensitivity to deflection is so greatly improved by use of the short focusing coil that its use is more than justified. With this kind of focusing, the plane of the electrons' focus is very deep, in fact an approximate focus exists between the plane *A—A'* (see Fig. 48) and the fluorescent screen, and close adjustment of the current in the focusing coil is not required. Also the beam stays in focus when deflected, which is a great advantage.

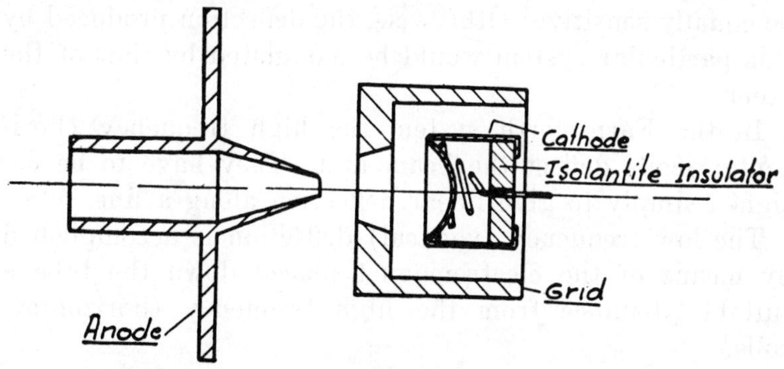

Fig. 49. The Electron Gun of the Oscillight Showing the "Protuberant" End on the Anode

The Deflecting Coil

Having focused the beam, the next problem is its deflection. The simplest system of deflection is that of an air-core coil positioned as close as possible to the walls of the oscillight tube and inside the focusing coil. This type of deflection system is perhaps the most sensitive that can be used, and is to be preferred, especially for use at frequencies of the order of 1,000 to 10,000 cycles per second.

Another type of deflecting system is the electromagnet shown in Fig. 48. The principal advantages of the use of the electromagnet is that the deflection is concentrated at a point along the tube and minimizes the interference which results into nearby circuits and deflection coils. The ferromagnetic type is preferable for the low frequency (vertical) deflection. The disadvantages in the iron-core type deflection system are that it is less sensitive than the air core, and that troublesome distortion results because of the hysteresis of the core material.

When it is attempted to put two deflection coil systems close together, deflecting at right angles to each other, several factors must be taken into consideration. Since the deflections take place simultaneously, it is necessary that all parts of the deflection field of one deflecting system be equally sensitive. Otherwise, the deflection produced by this particular system would be modulated by that of the other.

In the Farnsworth system the high frequency (horizontal) coils deflect the beam first. They have to be designed simply to give linear deflection along a line.

The low frequency (vertical) deflection is accomplished by means of the electromagnet placed down the tube a suitable distance from the high frequency (horizontal) coils.

These pole pieces must produce a uniform deflection no matter in what part of the tube the beam may be. To accomplish this requirement the pole pieces have been given the shape shown in Fig. 48, and this suffices if the pole pieces are placed the correct distance apart.

Mr. Farnsworth points out that it is entirely practical to deflect the beam electrostatically even though the focusing is obtained electromagnetically. The condition for satisfactory electrostatic deflection, however, is that the deflection plates must extend practically the full length of the focusing coil.

CHAPTER IV

THE FARNSWORTH RECEIVING SYSTEM COMPLETE

How the Oscillight Is Used

The Farnsworth receiving system centers itself about the cathode ray tube known as the oscillight, described in the previous chapter. Magnetic fields are employed for focusing and deflecting; and this feature offers two outstanding advantages: (*a*) simplicity and comparative cheapness of construction of the oscillight, and (*b*) sharpness of focus throughout the picture field—difficult to obtain by any other method.

It is obvious that any electronic television receiving set will have to be built around the special cathode ray tube, as it were, and that it is therefore desirable that this tube be as simple and inexpensive of construction as possible. This cathode ray tube will be the part that will wear out first and which will require replacement. At the present time it is probably rather premature to speak of the cathode ray receiving tubes wearing out, and the oscillight in the laboratory has given promise of exceedingly long life, yet this phase of television must be anticipated.

Again, in the future, when the amateur enters the electronic television field and begins to make his own receiving set, the cathode ray tube will be the essential part that he will have to first procure as the nucleus for his experiment.

The oscillight, using magnetic focusing, seems to solve these problems of simplicity and minimum of cost of construction, all within the limits of good performance.

Had electrostatic focusing been used, it would have been necessary to build deflecting plates in the tube, and this, naturally, would have added considerably to the cost of the tube itself.

Fig. 50. The Farnsworth Television Receiver for Television and Sound Mounted in Cabinet

The length of the oscillight is also an important feature. This length is such that it may be mounted on the chassis, and in the cabinet, in a horizontal position. It is therefore possible to view the pictures by looking directly at them

Fig. 51. Rear of the Cabinet of the Farnsworth Electronic Television Cabinet. On the upper shelf will be seen the Farnsworth reproducer unit for television. On the bottom shelf is the sound equipment

as they appear on the fluorescent screen of the oscillight. There is a psychological effect in this, since people are accustomed to look directly at pictures.

With longer tubes, it is most convenient to mount the

tube in the chassis and cabinet in a vertical position, because if the tube were mounted horizontally it would mean that the cabinet would require an inconvenient depth. Thus the picture appears on the upper surface of the cabinet,

Fig. 52. Baird Receiver (England)

and some means had to be devised to make the pictures visible to people sitting in a room. This problem was solved by placing a mirror on the inner side of the cover of the cabinet (see Figs. 52, 53), and with this mirror at an angle the picture is visible in it. However, as much as 10 per cent of the illumination is lost by this method.

It would therefore seem that, all other things being equal, it would be most desirable to have the cathode ray tube mounted horizontally.

The Farnsworth reproducer, possessing all of the essen-

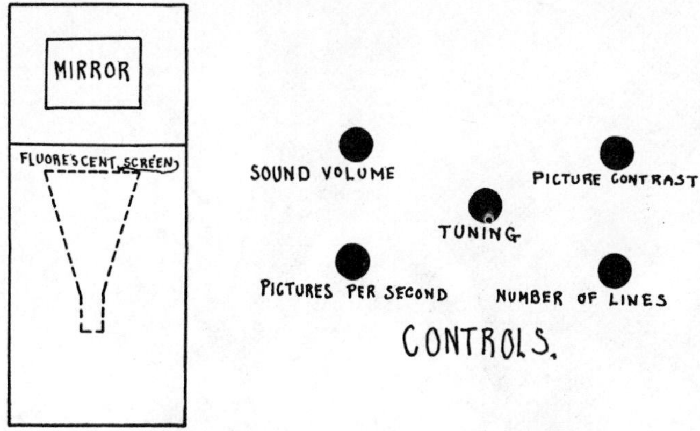

Fig. 53. Controls on Receiver shown above

tials of a television receiver (see Fig. 54), except the radio frequency and detector circuits, is compactly mounted on a chassis. This includes the magnetic circuits for focusing and scanning with the oscillight, oscillators to deflect the ray in scanning, and the high voltage supply for the anode of the tube.

The **chassis** includes (see Fig. 55):

A. A single amplifier stage for the picture signal, comparable to the output stage of a sound receiver except for its wide-band amplification. There is a gain of 100 in this amplifier.

Horizontal Scanning System

B. A tube which amplifies and detects the line synchronizing impulses which are transmitted with the picture signal itself. Thus the line synchronizing impulses are separated from the signal on the basis of their amplitude and polarity, and they are applied to the horizontal scanning oscillator to attain automatic synchronization.

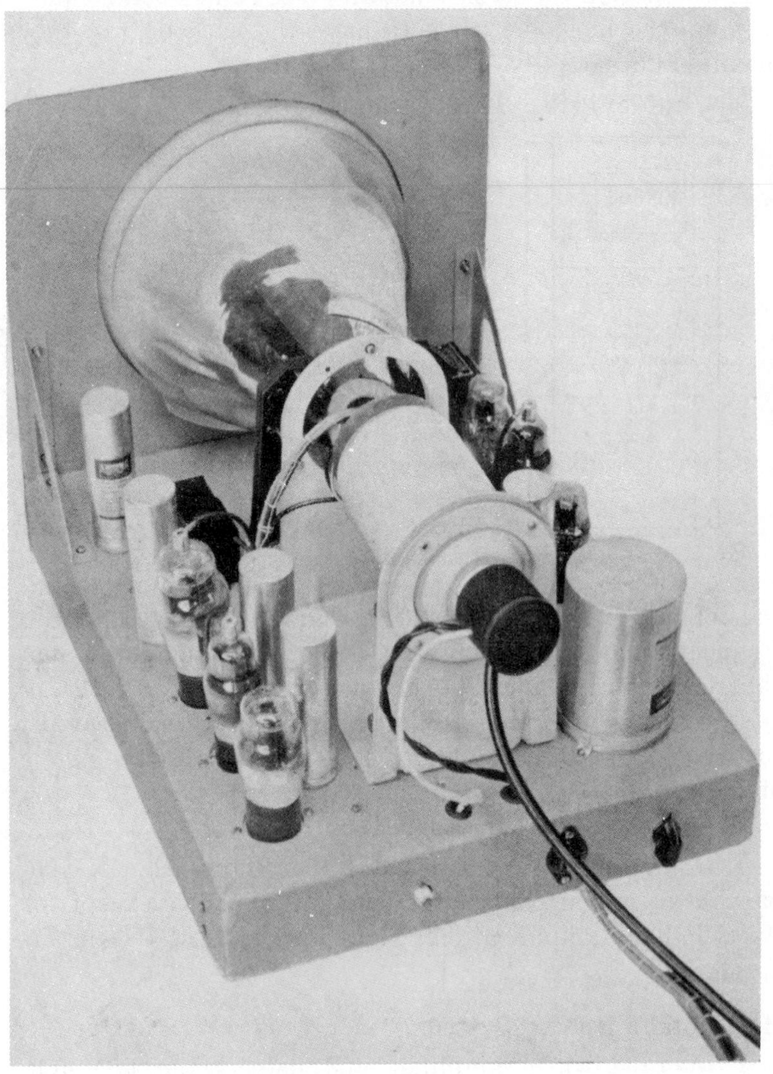

Fig. 54. The Farnsworth Reproducer Mounted on Chassis

C. A vacuum tube oscillator which supplies current of the proper frequency and of a sawtooth wave shape for

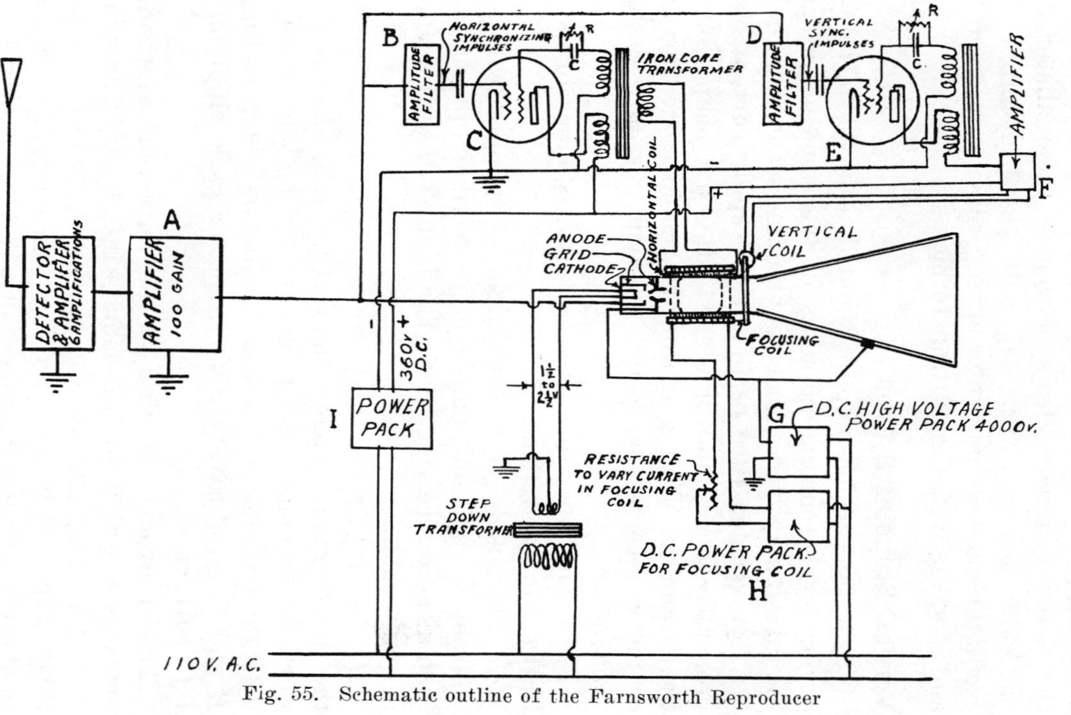

Fig. 55. Schematic outline of the Farnsworth Reproducer

scanning. The return of the scanning spot requires only about one-tenth the time consumed in scanning one line, and during this return period an impulse is received over the picture channel which synchronizes the oscillator, automatically holding it in step with the transmitter. The oscillator employs one tube of an ordinary receiver type and a specially designed transformer.

Vertical Scanning System

D. A tube which amplifies and detects the picture-frequency synchronizing impulses in the same manner that the tube *B* amplified and detects the line-scanning impulses.

E. A vacuum tube oscillator to supply current of the proper frequency and of a sawtooth wave shape for vertical scanning. This oscillator also employs one tube of an ordinary receiver type and a small specially designed transformer.

F. An amplifier tube as shown at *F* is added to the vertical scanning system.

G. The anode of the oscillight is supplied with 4,000 volts by a small unit employing a rectifier tube no larger than a radio receiving tube. The high voltage transformer is quite small.

H. The D.C. power pack for supplying D.C. current for the focusing coil employs one rectifier tube. In the Farnsworth System one power pack supplies the D.C. current for both the focusing coil and the oscillators. Thus in Fig. 55, *H* and *I* are, in reality, a single power pack supplying the D.C. current.

The electronic scanning in the Farnsworth reproducer operates on essentially the same principles as it does in the dissector. A special tetrode, similar to the ordinary power tubes employed in radio receivers except for the grid used in synchronizing is employed to supply the respective sawtooth scanning currents. The circuit is similar to that of the conventional vacuum tube oscillator, but the peculiar wave form results from two unusual circuit parameters:

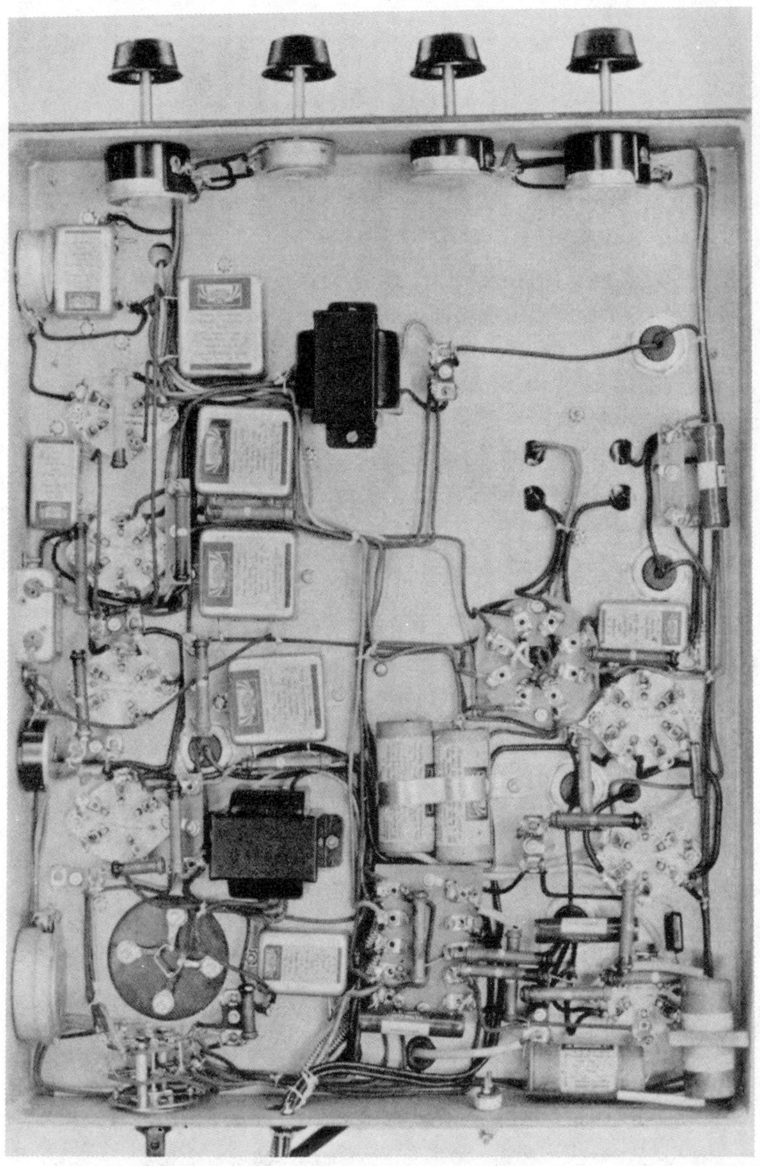

Fig. 56. Under Side of the Chassis of the Farnsworth Reproducer

Fig. 57. A German Television Receiver Made by Fernseh Akt.-Ges. This receiver is for television pictures and sound

(1) the iron-core transformer works in the region of saturation during part of the cycle, (2) the grid of the tube is over-excited by an excessively large voltage developed in the grid winding of the transformer. The frequency is adjusted by varying the time constant of the R.C. circuit in series with the grid winding (See Fig. 55). In the reproducer, when the frequency is set to the approximately correct value by variation of the grid leak, it is controlled by impulses applied to the synchronizing grid of the tube. The generators have been found entirely dependable.

The Farnsworth receiving system is compact and comparatively simple. The chassis carrying the reproducer can be placed in a cabinet of pleasing design, leaving ample space for the sound equipment. The cabinet can be so designed as to harmonize with other furnishings in a room, and if the picture appears on the front of the cabinet at a convenient height the television receiving set occupies little, or no more, space than a present-day console radio cabinet.

CHAPTER V

THE KINESCOPE

The "heart" of the R.C.A. Electronic Television Receiving System is the **kinescope**, a cathode ray tube of special design. In the kinescope the electron beam is focused electrostatically. This feature will be discussed later in the chapter, since it has an important bearing upon certain other cathode ray tubes used for purposes even outside of the realm of television.

The kinescope (see Figs. 58 and 59), is furnished with an electron gun made up as follows:

1. The indirectly heated cathode *C*, with its emitting area located at the tip of the cathode sleeve, and formed by the usual coating of barium and strontium oxides. The cathode *C* operates on alternating current.

2. The control electrode *G*, corresponding to the grid in the ordinary triode. It is, of course, by means of this grid that the incoming picture signal modulates the electron beam emitted by the cathode *C*. The control element *G* has an aperture, *O*, directly in front of the emitting surface of the cathode *C*, and besides acting as the control element also serves as a shield for the cathode. This shield, which is furnished with a negative voltage, sets up an electrostatic field which repels the negatively charged electrons, coming off from the heated cathode *C* in a "mist," in such a manner that they are "crowded" toward the hole in the first anode, A_1.

3. The first anode A_1 has suitable apertures which limit the angle of the emerging electron beam. The high positive voltage of the first anode A_1 accelerates the beam.

It is obvious, however, that the electron beam which is to finally strike the fluorescent screen must be sharply focused, in fact the cross-section of the beam should be of the same area as an element in the pick-up in order to give definition in the reproducer as high as that in the pick-up.

Fig. 58. The Kinescope

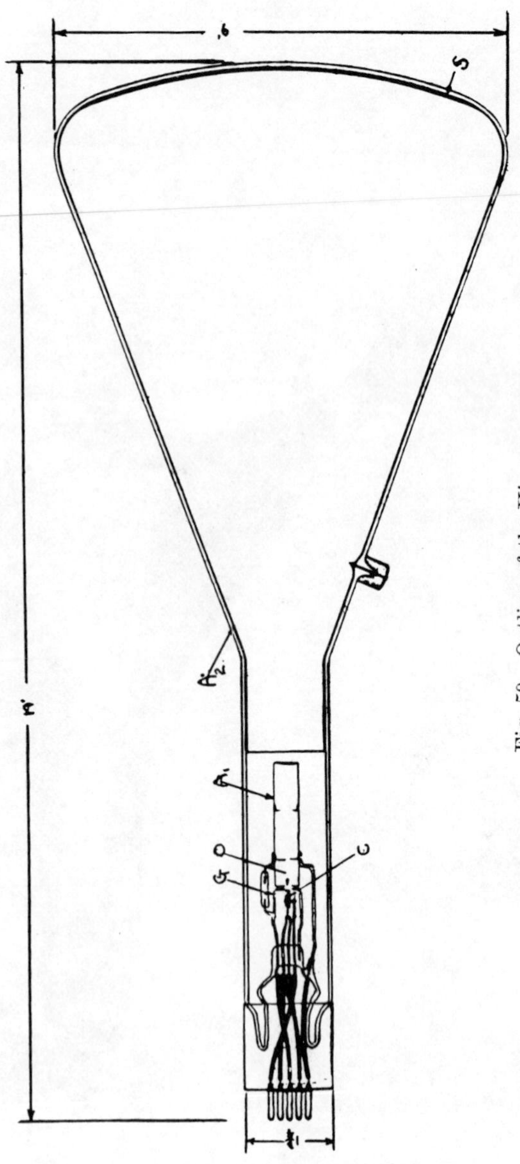

Fig. 59. Outline of the Kinescope

To attain this sharp focus the second anode A_2 is added. The inner surface of the cone of the kinescope is silvered, or otherwise metallized, and serves as the second anode, A_2. The purpose of the second anode is twofold: (1) it accelerates the electrons emerging from the gun, (2) it forms an electrostatic field which focuses the electrons emerging from the gun into a small, "thread-like" beam.

Electrostatic Focusing

Magnetic focusing of the electron beam is somewhat complicated by the presence of a strong electrostatic field which is used to increase the energy of the beam, or "to accelerate" the electrons of which the beam is formed toward the fluorescent screen. Therefore it suggested itself to Dr. V. K. Zworykin that it would be advantageous to use the electrostatic field set up by the anodes for both acceleration and focusing in the kinescope.

The emitted electrons have a natural tendency to diverge. The lines of force of the electrostatic field, between properly shaped electrodes, tend to force the electrons toward the axis of the beam, thus overcoming this tendency.

This action is very much similar to the focusing of light rays by means of optical lenses. There is an important difference, however, since electrostatic lenses have a peculiarity in that their index of refraction for the electrons is not confined between the optical media, as in optics, but varies throughout all of the length of the electrostatic field.

This can best be understood by studying Fig. 60, which shows the electrostatic lenses, and below the optical analogies. While the optical analogy is not quite correct, it will, however, serve the purpose of presenting the subject in the simplest manner.

It was found to be practically impossible to produce a simple single electron lens, for the field always forms a combination analogous to positive and negative optical lenses. However, by proper arrangement of electrodes and potentials it is always possible to produce a complex "electrostatic lens'' which will be equivalent to either the negative or positive optical lenses.

Again this is best understood by referring to Fig. 60, which shows the distribution of the electrostatic fields in the kinescope. This is equivalent to a combination of four optical lenses, as shown.

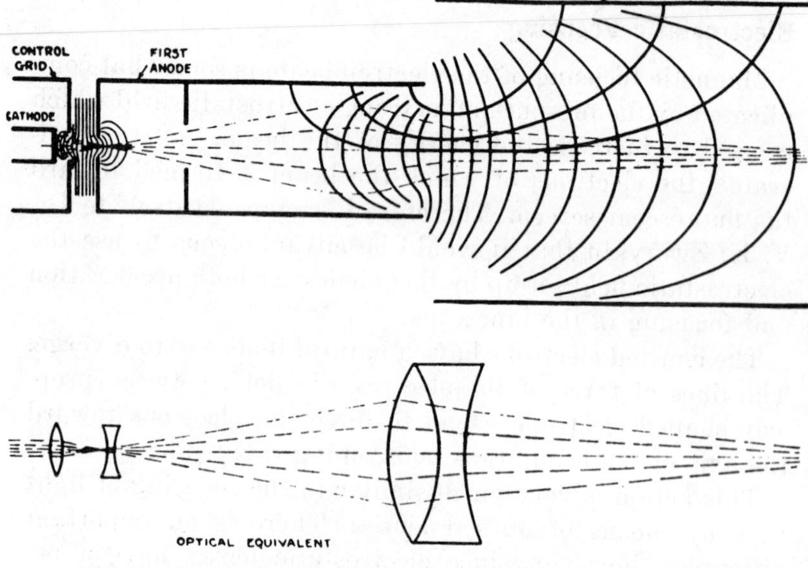

Fig. 60. Optical Analogy to Electrostatic Focusing in the Kinescope

In the kinescope, the first two lenses force the electrons through the aperture of the first anode, and assure control of the beam by the control element.

The final focusing of the beam on the fluorescent screen is accomplished by the second pair of lenses created by the electrostatic field between the end of the electron gun and the neck of the tube.

The first anode operates at a fraction of the second anode voltage.

Thus the final size of the spot on the screen, as in the optical analogy, depends upon (*a*) the size of the active area of the cathode, (*b*) the optical distance between the cathode, lenses, and screen.

The Fluorescent Screen

The fluorescent screen (*S*) (Fig. 59) is deposited upon the flat end of the cone of the kinescope. This fluorescent screen absorbs electrical energy and emits light, and is very thin so that a large part of the emitted light is transmitted outside of the tube as useful illumination. Material most used for the screen is synthetic zinc orthosillicate almost identical with natural willemite. This gives a light green color.

It is necessary to vary the intensity of the spot of light upon the fluorescent screen in order to reproduce the varying light intensities of the original picture, and this is accomplished by means of the control grid *G* of the electron gun.

For satisfactory reproduction it is obvious that the control of the electron beam intensity should be directly proportional to the input signal voltage. Or in other words, if the input signal voltage is plotted against the candle power of the resulting light intensity on the screen, the graph should be a straight line.

The "time decay" curve of the fluorescent screen material is also a matter of importance. It will be remembered that the fluorescent screen absorbs electrical energy and gives off light. Now this phenomenon is not absolutely instantaneous in any fluorescent substance; there is a "luminescence" that trails the electrical excitation. In other words, the screen "holds" the picture for a slight fraction of time after the beam bombardment at that point has ceased. This is known as time decay, the time that the particular fluorescent material "holds" the picture before it fades out after excitation has ceased.

Now this time decay has both advantage and a disadvantage. The disadvantage is that if the time decay is too long moving portions of the picture will trail. For instance, the picture of a baseball would carry a comet-like tail, for the pictures of the ball in various positions would still show on the screen while the next position of its flight were shown. If the time decay is too short, however, the pictures will flicker because of the periods of comparative total darkness between excitations of successive pictures.

On the other hand, each picture frame is scanned in the "pick-up," and reproduced in the receiver, by starting at the top of the picture frame and scanning downward. Therefore there is an advantage in having the top of the picture on the screen "hold" until the entire frame has been scanned. It is obvious that the time decay of the material from which the fluorescent screen is made must be just right, neither too long nor too short.

In interlaced scanning an interesting feature arises. The frame is scanned once, and then scanned again, the second set of horizontal lines falling between the first set. Thus, it will be seen, that there is "retouching" of the picture on the fluorescent screen, as it were; the first set of lines are beginning to "decay," when the second set, between the first, appear.

When the electron beam strikes the fluorescent screen, the bombardment of the negatively charged electrons would give the screen a negative charge, and because the material of which the screen is made is a good dielectric, this charge would remain on the surface and accumulate until the negative charge would be sufficient to completely repulse the beam from the screen, and if this happened, the light emission would, of course, cease.

At first a thin metallic screen was placed between the end of the kinescope, and the fluorescent screen, to carry off this charge. However, it was soon found that secondary emissions were bombarded out by the primary electrons of the beam. These secondary emissions were attracted to the highly positively charged metallized inner surface of the kinescope, the second anode, and were carried off. Thus it was found that the metallic screen was unnecessary.

CHAPTER VI

THE R.C.A. RECEIVING SYSTEM

How the Kinescope Is Used

It will be remembered that the problem of picture transmission is essentially three dimensional, two dimensions, height and breadth of the picture frame, to give area and a third to indicate varying intensity. A radio channel, of course, is only two dimensional, as it were, capable of transmitting intensity of signal and duration of time. This brings back the problem of scanning, the reassembling of the scene in the receiver and reproducer that had been taken apart in the pick-up.

The kinescope is capable of taking care of the third of the three-dimensional requirements stated above, namely the indication of varying intensity; and, therefore, horizontal and vertical deflection must be added to the kinescope in order to attain the essentials of a reproducer. Both the horizontal and vertical deflection frequencies must be the same, respectively, as those in the pick-up, and must be in synchronization with them.

The Deflecting Coils

In the R.C.A. Electronic Television Reproducer the deflection is attained electromagnetically. This choice was the result more of economical consideration than technical. It must be remembered that electrostatic deflection is possible. However, it is much cheaper to make a kinescope for electromagnetic deflection than one for electrostatic deflection, since in the latter case inside deflecting plates must be built in the tube.

Schematically the R.C.A. receiver and reproducer is as shown in Fig. 61. The incoming signal is made up of (*a*) the picture signal proper, (*b*) the horizontal synchronizing impulses, (*c*) the vertical synchronizing impulses.

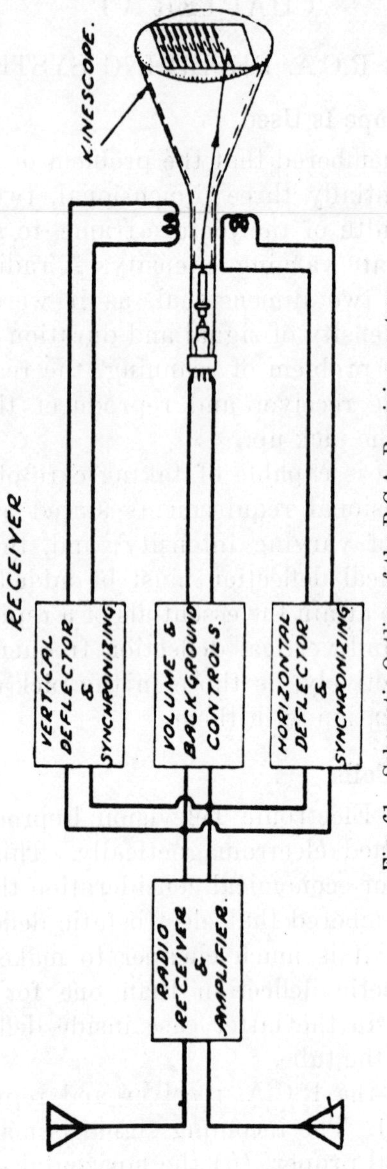

Fig. 61. Schematic Outline of the R.C.A. Receiving system

By amplitude and frequency selection the horizontal and vertical scanning impulses go to the respective generators. The entire incoming signal goes to the kinescope, where the synchronizing impulses serve to extinguish the electron beam during the back traces in both horizontal and vertical scanning.

The synchronizing impulses are in the negative direction, and when these are applied to the control grid of the kinescope they bias the grid negatively and extinguish the beam.

The picture background, or average illumination of the picture, can be controlled by adjusting the kinescope bias.

The problem of horizontal and vertical deflection currents

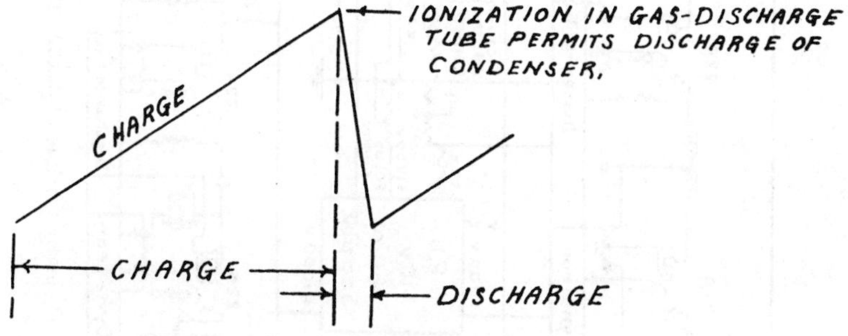

Fig. 62. Simple Method of Generating Sawtooth Wave Current

to furnish the respective magnetic fields brings up again the subject of the generation of sawtooth wave currents.

There are many ways of generating sawtooth wave impulses. Probably the most simple method consists in charging a condenser through a current limiting device, such as a saturated two-electrode vacuum tube, and then discharging the condenser through a thermonic or gas-discharge tube.

Briefly this is as follows: the two-electrode vacuum tube is made to operate in a saturated manner. Thus it charges the condenser at a steady rate. When a certain predetermined voltage is reached ionization occurs in the gas

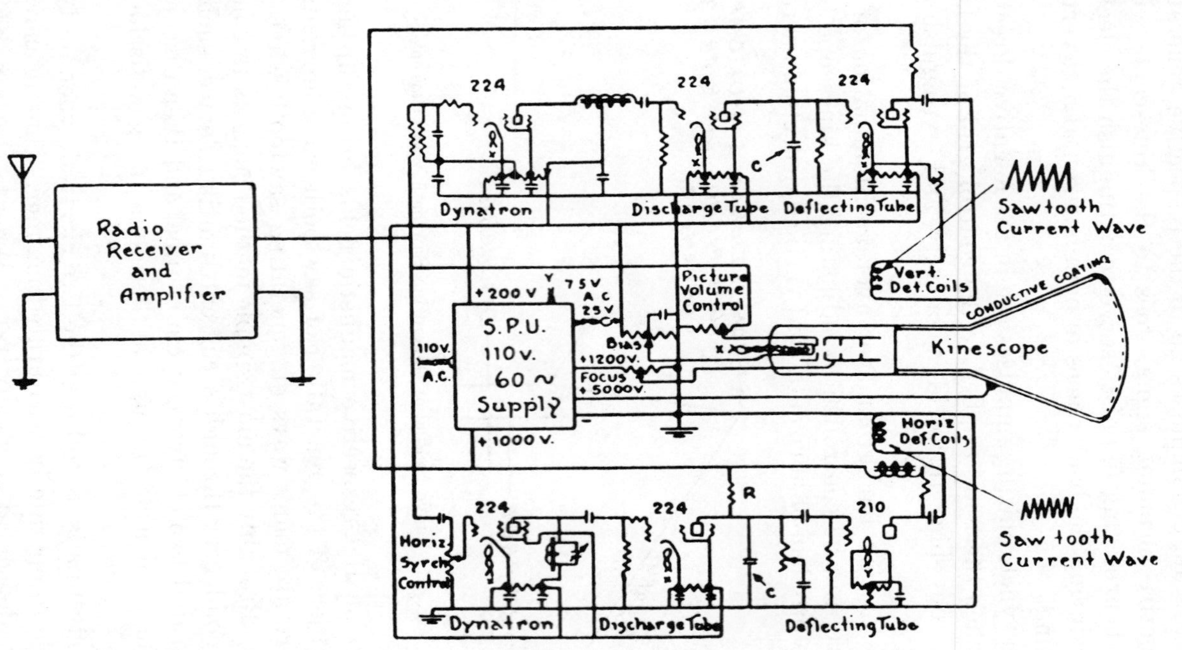

Fig. 63. Schematic Outline of R.C.A. Receiver

discharge tube, and permits the condenser to discharge almost immediately down to a voltage where ionization is no longer possible, and then the process begins all over again.

While theoretically this sawtooth wave generator is quite simple, it is, in practice, limited by the fact that there

Fig. 64. The R.C.A. Receiver

is no such thing as a completely saturated thermionic tube. Therefore the line of charge in Fig. 62 would not be exactly straight, and this, of course, would have an effect on the resultant scanning line on the fluorescent screen. Therefore a more complicated circuit was used (see Fig. 63).

This circuit consists of one dynatron oscillator and two amplifying tubes. In the horizontal deflecting circuit the condenser C is charged continuously through the resistance R.

Periodically, at the end of predetermined intervals, the condenser C is discharged. During these time intervals the accumulated charge does not reach saturation, for the time is insufficient. It must be remembered that in the horizontal de-

Fig. 65. Rear View of R.C.A. Receiver. Note the Kinescope placed vertically

flection system this time in seconds is (*one*) over (*the number of lines per picture* multiplied by *the picture frequency*).

The vacuum tube through which the discharge takes place is controlled by impulses supplied from a dynatron oscillator having a distorted wave shape. The frequency of oscillation of the dynatron which can vary over quite a large range is at first adjusted to a frequency approximately that of the scanning frequency. The synchronizing impulses then have no difficulty in pulling the dynatron into "step" with the pick-up scanning.

The charging and discharging of the condenser *C* represent sawtooth variations of potential, which, when applied to the grid of an amplifying tube, produce sawtooth wave current impulses.

The R.C.A. Receiver

In the R.C.A. receiver the kinescope is placed vertically. The chassis slips into the cabinet as a unit and consists of a power unit, kinescope unit, two radio receivers, one for picture and one for sound, and a loud speaker.

The reproduced image is viewed in a mirror mounted on the inside lid of the cabinet. This shields the picture from overhead illumination, and affords a convenient viewing angle. The brilliancy of the picture is sufficient to permit observation without the necessity of completely darkening the room.

PART III

THE BY-PRODUCTS OF ELECTRONIC TELEVISION RESEARCH

CHAPTER I

SECONDARY ELECTRON MULTIPLICATION AND MULTIPACTOR TUBES

Secondary electron multiplication and multipactor tubes were discussed in chapter III, Part I, in the explanation of the Farnsworth Image Dissector Tube. The subject is of such great importance to the future of the science of radio communication, even beyond the field of electronic television, that it seems well to go into it more fully.

Farnsworth Multipactor Tubes have been designed and made to be used as amplifiers, oscillators, frequency multipliers, detectors, in short, wherever a radio tube is now used (see Fig. 66). These tubes have been of various types, some with, and some without, filaments, according to the use for which the tube was designed.

This new type of tube is based upon the harnessing of secondary electron emission properties of metals. It has long been known to scientists that metals have electrons in suspension upon their surfaces, which may be released when "bombarded" by other electrons.

By taking advantage of this effect, an effect studiously avoided and considered detrimental by the designers of thermionic (hot) radio tubes, Mr. Farnsworth evolved a radically new tube which operates with greatly increased efficiency and eliminates many of the disadvantages of the tubes now used for broadcasting and receiving. It must be remembered that the multipactor tube is essentially a "cold" tube, as distinguished from the thermionic or "hot" tube.

This hitherto avoided effect of secondary electron emissions resulted in decreasing the output and efficiency in previously designed thermionic tubes, and, as a matter of fact, engineers were hard put to find ways of eliminating it. As a result the elements of many of the present tubes

Fig. 66. Small Radio Transmitter Using Farnsworth Multipactor Oscillator on Low Power

are purposely coated with graphite, or carbonized, so that secondary electrons will not be given off when the primary electrons from their hot cathodes strike the surfaces of their elements.

Probably the most interesting feature to the radio engineer is that in some of the designs of the Farnsworth Multipactor Tube no filament, or heated cathode, of any type is used or needed. The great advantage of this can be seen when it is remembered that the life of ordinary tubes now in use ends when the filament burns out. Therefore, if these filamentless Farnsworth Multipactor Tubes are operated within their ratings, there appears to be no good reason why they should not last indefinitely.

The basic theory upon which the Farnsworth Multipactor Tube works, and which was briefly discussed in chapter III, Part I, is extremely simple, and the manner in which Mr. Farnsworth used this theory is quite ingenious.

In the filamentless Farnsworth tubes the few stray electrons which are always present due to photoelectric and cosmic ray effects are bombarded against a surface especially prepared to have the best possible secondary emitting properties.

As a result, these original electrons, on striking this surface, cause emissions of many times more electrons than were in the original bombardment. This procedure is repeated many times until the desired "electron amplification" is reached.

However, due to the special design of the Farnsworth Multipactor Tube, the process is controlled with great precision. This electron amplification builds up in such extremely short-time intervals that, if allowed to go unchecked, the tremendous current produced would fuse the elements of the tube. By application of this multipactor principle, Mr. Farnsworth has also increased the output and efficiency of his own filament type tubes, resulting in greater output current and greater amplification than would be at all possible without taking advantage of secondary emissions.

In chapter III, Part I, one type of multipactor, the radio frequency type of multiplier, was discussed. The *constant potential type of multiplier* is another example of how the process of secondary electron multiplication may be carried out practically.

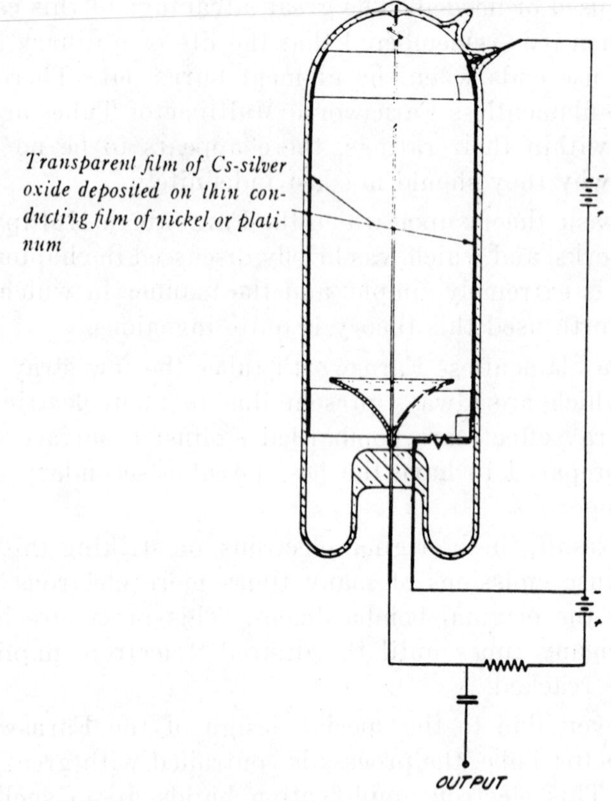

Fig. 67. Diagram of Farnsworth Electron-Multiplying Photoelectric Cell

Constant Potential Type Multiplier

Fig. 67 represents an electron-multiplying photoelectric cell, designed by Mr. Farnsworth, in which the photoelectric currents from the upper end of the tube may be multiplied many times by successive secondary emission impacts.

Briefly the tube is constructed by evaporating on the walls of a tubular bulb a thin film of platinum or nickel, and making connection at each end of this film by means of a platinum tab fused in the glass. On top of the nickel film is evaporated a film of pure silver to a thickness that is nearly opaque.

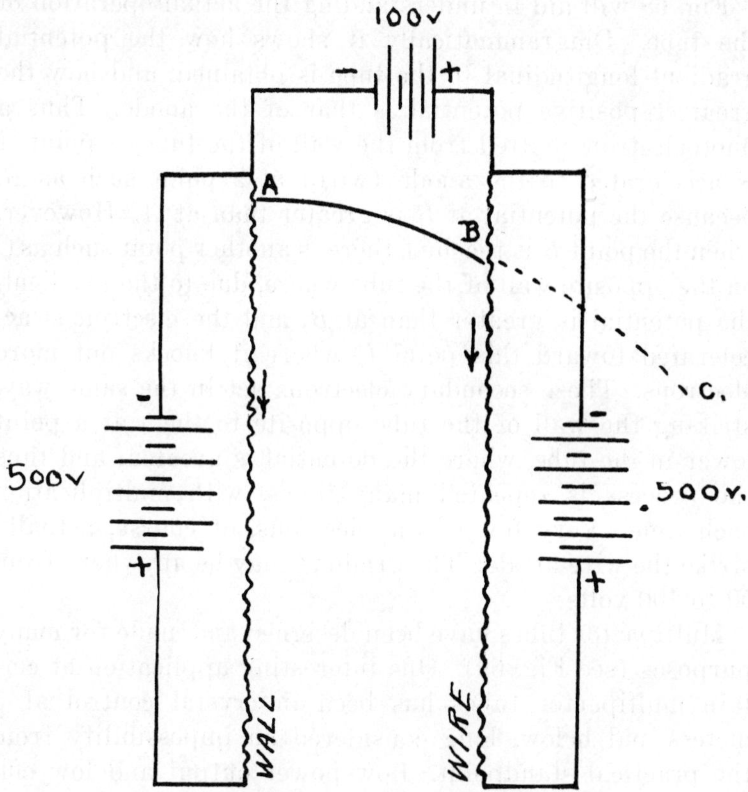

Fig. 68. Diagram of Potential Gradient

Down the axis of the tube is mounted an anode consisting of a fine wire or loop of wire, having at one end a metallic collection plate as shown.

The tube is sealed onto the pump, pure oxygen is admitted, and the silver surface completely oxidized. Caesium is then driven into the tube, and the tube formed into a photoelectric cell. After the photoelectric surface has been formed, the tube is sealed off the pump.

In use, a potential of 500 volts or more, is applied across the distributed resistance so that a potential gradient longitudinal to the tube results. At the same time the greatest positive potential in the tube is that of the anode.

Fig. 68 will aid in understanding the actual operation of the tube. Diagrammatically it shows how the potential gradient longitudinal to the tube is obtained, and how the greatest positive potential is that of the anode. Thus a photoelectron emitted from the wall of the tube at point *A* is accelerated to the anode (wire) at a point such as *B*, because the potential at *B* is greater than at *A*. However, when the point *B* is reached, there is another point such as *C* on the opposite wall of the tube where, due to the gradient, the potential is greater than at *B*, and the electron is accelerated toward this point *C* where it knocks out more electrons. These secondary electrons act in the same way, striking the wall of the tube opposite to them at a point lower in the tube, where the potential is greater; and thus the process is repeated many times, with multiplication each time. Very few of the electrons, of course, actually strike the wire anode. The gradient may be anywhere from 50 to 100 volts.

Multipactor tubes have been designed and made for many purposes (see Fig. 69). One interesting application of certain multipactor tubes has been in crystal control at 5 meters and below, long considered an impossibility from the practical standpoint. Low power output and low efficiency, coupled with numerous technical problems, have forced the amateur to turn to other means of frequency control. Crystal control of frequencies of from 50 to 150 megacycles has been achieved in circuits making use of the multipactor. The multipactor, it must be remembered, is primarily an electron multiplier in which the principle of

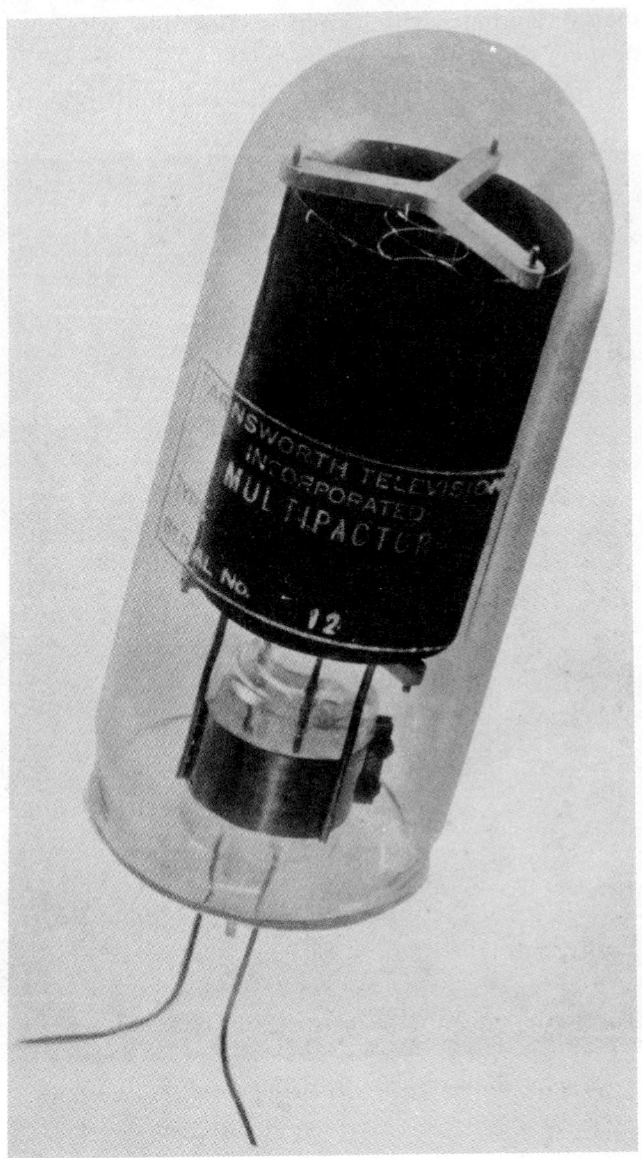

Fig. 69. Farnsworth Multipactor Tube

secondary emission produces both high efficiency and rich harmonic content which makes possible ultra-high frequency crystal control.

The principle of secondary electron multiplication will

Fig. 70. Dr. Zworykin with His Electron Multiplier Tube

undoubtedly have a far-reaching effect on future developments in all branches of radio science, including television. It is well within the realm of possibility to predict that these tubes will emerge from the laboratory and, becoming available to amateurs and engineers, will possibly revolutionize radio design.

Mr. Farnsworth has suggested that the multiplication principle may be applied to many uses besides amplifying an electron flow, such as: (*a*) Image amplifier and translation device for such uses as an infra-red camera for fog penetration, electrical microscope or telescope, and general intensification of optical images. (*b*) Generator of oscillations of any frequency, particularly adapted for frequencies above 20 megacycles. (*c*) Radio frequency amplification for transmitter use. The output may be readily modulated. (*d*) Short-wave radio receivers.

Dr. V. K. Zworykin, of the R.C.A. laboratories, has also pursued intensive research in the realm of secondary electron multiplication (see Fig. 70) and electron multiplier tubes, and most interesting developments may be looked forward to from this source also.

CHAPTER II

INFRA-RED CAMERAS FOR FOG PENETRATION AND ELECTRON MICROSCOPES

Despite the fact that electronic television itself is just emerging from the laboratory stage, the by-products, as it were, of research in this field are already appearing, and these, in themselves, promise startling and wonderful developments.

In the first place, it must be remembered that certain surfaces, such as caesium-silver oxide surfaces, emit photo-electrons under the influence of light. In fact this is one of the fundamental principles behind all electronic television. However, these emissions are not limited to influences from radiations from light in the spectrum visible to man—the well-known band from red to violet—but extend into the ultra-violet and infra-red, the bands which the eyes of man cannot see.

Referring to Fig. 71, one curve shows the relative response of a caesium-silver oxide surface to radiations of light waves of various lengths, that is, in the various color bands of the spectrum. It must be remembered that this curve, as shown, is merely to give a general idea of the response, and is not a specific curve. However, it will be seen that there is response in both the infra-red and ultra-violet.

Again, a second curve on Fig. 71 shows the spectral sensitivity of the human eye. It must be repeated that these curves are not drawn to scale, but are general. The human eye is more sensitive to some colors than to others. A third curve shows the spectral sensitivity of zinc ortho-silicate or willemite used on the fluorescent screen. It will be seen that the pale green light produced by bombardment of the willemite screen falls near the color range to which the human eye responds most strongly.

Now with these curves in mind, it is obvious that, if through some device, emissions produced by the influence of infra-red, or "black light" could be bombarded against a screen of willemite, and the emitted electrons kept in "focus," an invisible image, that is, one invisible to the human eye, could be reproduced so that the human eye could see it.

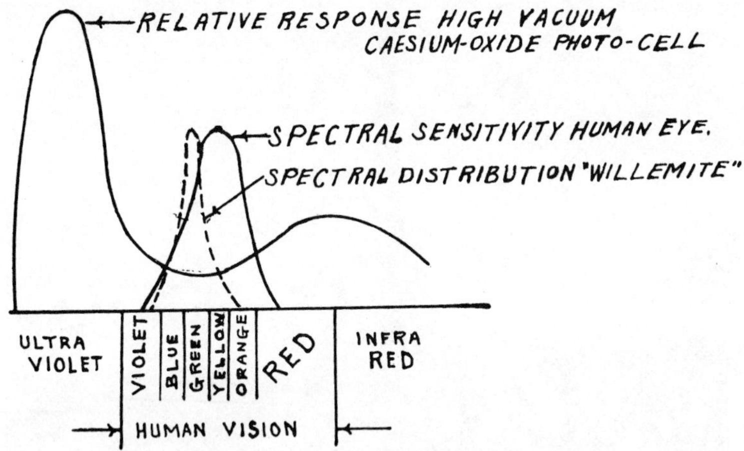

Fig. 71. Response and Sensitivity of Photo-Cell Compared to Eyes of Man

Infra-Red Penetration

In chapter II, Part I, it was shown how an electron image was produced when an optical image was projected upon a caesium-silver oxide surface. At every point of the surface electrons were emitted, and these emissions were in proportion to the amount of light falling upon that point. However, the electrons leave the surface in all directions, and at an infinite distance from the caesium-silver oxide surface the electron image would appear as a random cloud were they not focused. In the Farnsworth Image Dissector Tube they are focused electromagnetically, it will be remembered.

In chapter V, Part II, it was shown how Dr. V. K Zworykin focused the beam of the kinescope electrostatically.

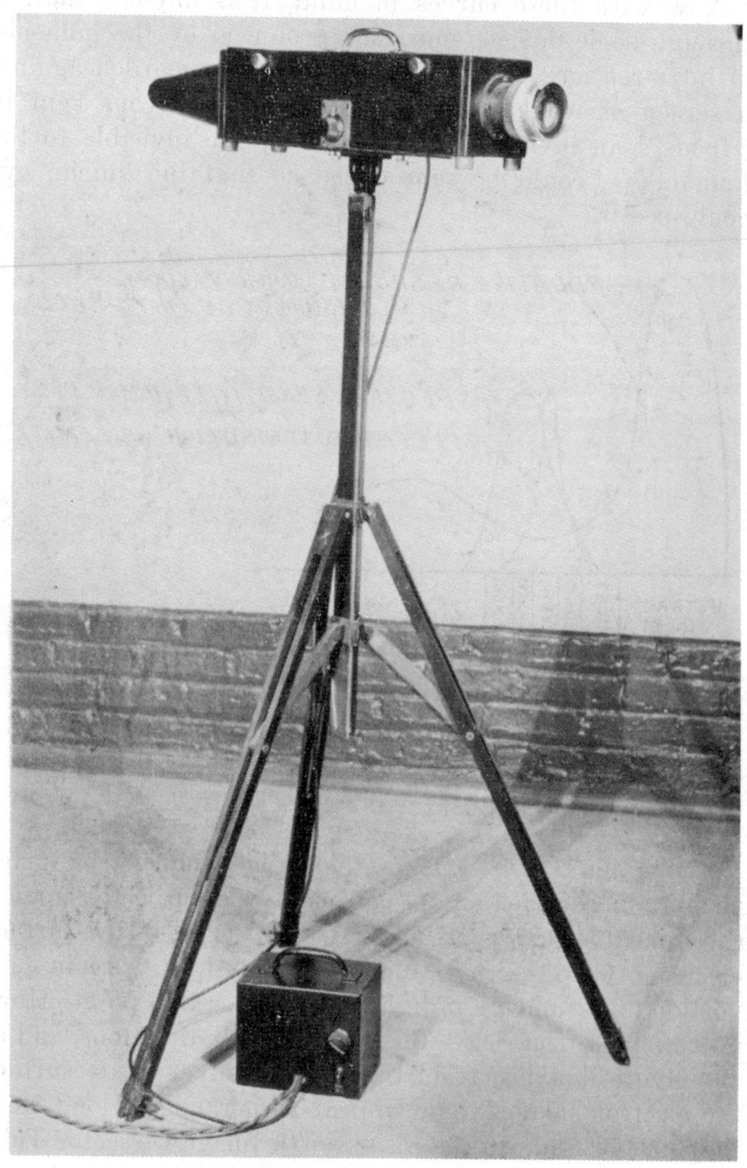

Fig. 72. Electron Telescope

This touched upon electron optics, a comparatively recent branch of the science of electronics, based upon the similarity of electron paths through certain types of electric fields and the paths of light rays through ordinary lenses.

Dr. Zworykin pointed out that this field of study shows

Fig. 73. A Scene Projected in Infra-red on the Fluorescent Screen of the Image Tube

that it is possible to shape electrodes in such a way that the electric fields between them will act as an "electron lens," capable of focusing the electrons leaving the cathode into an image of that cathode. Such an electron lens is found to have properties almost identical with those of an ordinary glass lens.

Fig. 74. Image Tube and Microscope

For example, Dr. Zworykin again pointed out that the image must be in focus if it is to be sharp; the magnification is determined by the distance from electron source or object to the lens and from lens to the object, and furthermore the image will be inverted as is the case with an optical image. Just as the camera lens must be corrected

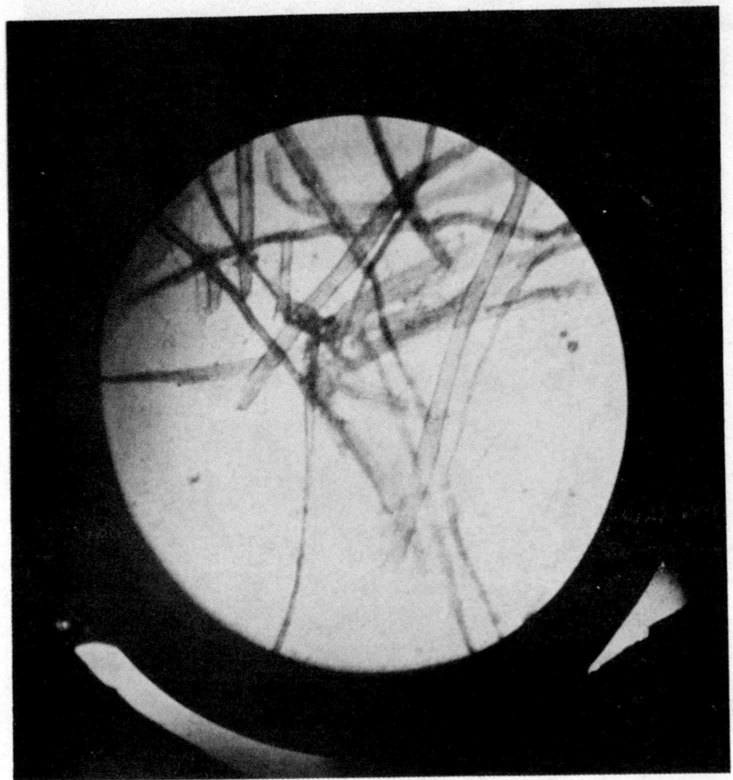

Fig. 75. Visible Image on the Fluorescent Screen of a Micro Specimen

if the image is to be free from distortion, so must the electron lens be corrected in order to obtain a perfect image.

The electron image tube developed by Dr. Zworykin may first be likened to a complete electronic television system in one tube, with the pick-up on one end, and the reproducer showing a picture by bombarding an electron image, in focus, against a willemite screen on the other end. Of course, it is not as simple as all this, but these are the essentials, and not only do the electron lenses focus the electron image, but they are also capable of producing magnification.

Fig. 76. Image Distortion in Uncorrected Image Tube

Fog Penetration

Thus it will be seen that such a device can be used to test haze and smoke penetration by infra-red, and that objects may be made visible in ''black light'' (see Fig. 73).

Another use is in connection with infra-red microscopy. Fig. 74 shows an image tube and microscope arranged for infra-red work. By means of this device, sensitive to the infra-red rays, it is foreseen by some that the developments of hitherto baffling minute living organisms may be brought within the range of human vision, since such cells have been studied in the past by means of intense light or stains that often kill them (see Fig. 75).

Fig. 77. Undistorted Image in Corrected Image Tube

In the R.C.A. Electron Image Tube developed by Dr. Zworykin, a high positive potential is supplied to the long anode cylinder to accelerate to a high velocity in order to produce fluorescence on the screen. The anode, together with the focusing rings, form a "lens" system which can be adjusted by varying a second potential. The lens system will image a rectangular grid into an image such as shown in Fig. 76. In order to correct for the "pin cushion" distortion and curvature of the image field, the cathode is curved, and Fig. 77 shows the finally corrected image obtained.

In order to be able to vary the magnification of the lens system, an aperture is introduced between the anode cylinder and the focusing rings. Varying the potential of this aperture in effect "shifts" the position of the lens, causing a variation in magnification.

Electron Microscope

Research has also been carried on in the ultra-violet field. Philo T. Farnsworth and others have also experimented with the possibilities of the electron microscope.

It is not without the realm of possibility that some day the electron microscope will prove of tremendous service in the field of medical research.

Interesting indeed are the possibilities in the ultra-violet field. It is at once apparent that no organism smaller than a wave-length of visible light could be seen in any present-day optical microscope no matter how powerful. Now the ultra-violet rays are shorter than any visible rays of light. Therefore, with an electron microscope, sensitive to invisible ultra-violet rays, it might be possible, in fact it would be possible, to reproduce the electron emissions resulting from these invisible rays, upon the screen of an electron microscope. Thus it would be possible to actually "see" organisms smaller than a wave-length of visible light. It is possible that organisms causing certain diseases have defied identification and detection by medical men because those organisms are smaller than a wave-length of visible light. This opens a wide medical possibility for the electron microscope of the future.

CHAPTER III

DISTANCE ATTAINED IN ULTRA-HIGH FREQUENCY TRANSMISSION

Engineers have looked longingly toward the ultra-short wave portion of the radio spectrum, since the tremendous bands there available promise to furnish the vehicle needed for new radio services. A radio channel for television of good definition, for example, might require a band of radio frequencies two million cycles wide. When it is considered that present broadcasting channels are only ten thousand cycles wide, it is apparent that new radio services must find their vehicle of transmission in new and hitherto unused portions of the radio spectrum. Such an area is found in the realm of ultra-high frequencies or, to state it inversely, ultra-short waves, which comprise that part of the radio spectrum between ten meters and one meter.

Not only are wide channels necessary for the transmission of electronic television of high definition, but these wave bands are useful for the transmission of facsimile, and also afford many communication channels. There was one tremendous stumbling-block, however—the transmission range of ultra-high frequencies (ultra-short waves) has heretofore been limited to "line-of-sight" distances. In short, the effective distance of ultra-high frequency transmission has been, in the past, limited to the optical horizon; and by placing the sending antenna upon a tall building and having the receiving antenna placed high, about 50 miles was regarded as the effective range.

The R.C.A. Three-Meter Radio Circuit

This obstacle, the obstacle of distance, has now been overcome. The successful demonstration of a two-way ultra-high frequency circuit between New York and Philadelphia (See Fig. 78), operating on three meters, and using automatic, unattended relay stations, would seem to indicate that ultra-high frequency transmission no longer suffers the limitations of distance.

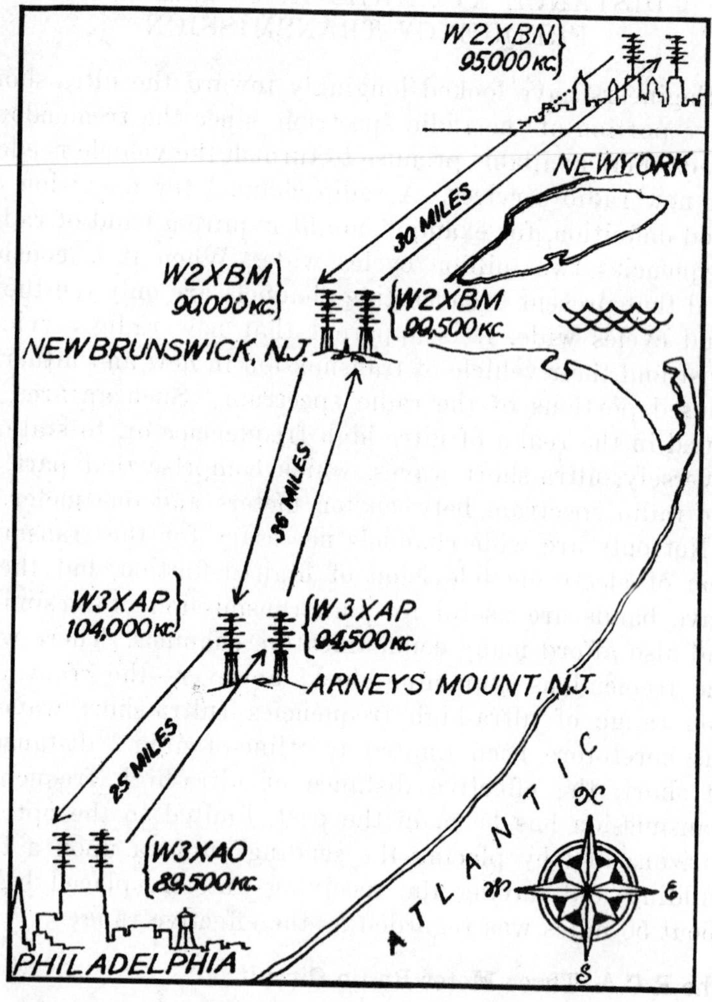

Fig. 78. Schematic Diagram of the R.C.A. New York-Philadelphia Ultra-High Frequency Circuit

Most ingeniously the research forces of the Radio Corporation of America have attacked and solved this previously baffling problem. It was, of course, apparent almost from the first that the distance could be extended indefinitely by the use of relay stations, but the operation and maintenance of a number of relay stations, manned by operators, would

Fig. 79. Transmitting Antennas Used on the New York End of R.C.A.'s New Ultra-Short Wave Radio Circuit to Philadelphia

be costly and cumbersome. The problem, therefore, that presented itself was to have automatic, unattended, relay stations that would not interfere in their reception and sending. This has been accomplished.

Briefly the R.C.A. New York-Philadelphia circuit is made

up as follows: In sending from New York to Philadelphia (southward): the New York station (W2XBN) (See Fig. 79), the antennas of which are placed on top of a tall building some 600 feet above sea-level, sends on a frequency of

Fig. 80. Antenna Atop an Office Building in New York for Reception of Ultra-Short Waves, in the New York-Philadelphia Circuit of R.C.A.

95,000 kc. At New Brunswick, N. J., 30 miles distant, relay Station W2XBM, with antennas some 250 feet tall, picks up station W2XBN, and retransmits at a frequency of 90,000 kc. This, in turn, is picked up by relay station

W3XAP, situated on a hill at Arney's Mount, N. J., 36 miles distant, which in turn retransmits at 104,000 kc.

Twenty-five miles distant, atop a tall building in Phila-

Fig. 81. Three Meter Transmitter Used in R.C.A.'s New Ultra-Short Wave Radio Circuit Connecting New York and Philadelphia

delphia, is station W3XAO, which receives from relay station W3XAP.

Now in transmissions northward, from Philadelphia to

New York, Station W3XAO, Philadelphia, sends at a frequency of 89,500 kc. This is picked up by Station W3XAP (Arney's Mount), and retransmitted at a frequency of 94,500 kc. which in turn is picked up by W2XBM (New Brunswick) and in turn retransmitted to New York at a frequency of 99,500 kc.

It will be seen that each of the relay stations employs two different transmitting wave-lengths, one for sending southward and one for sending northward. The two termial stations each use a different wave-length, making a total of six wave-lengths, or frequencies used. These frequencies have been so ingeniously arranged that there is no interference, between transmission northward and southward, at the same time, from the same station.

Should it be desired to extend this circuit farther, these same six wave-lengths could be used over and over again in the same sequence. In short, two waves of the same length would be generated at points about one hundred miles apart, and would not interfere with each other, because of the "line-of-sight" limitation to their respective ranges. There seems to be no serious obstacle against extending a circuit such as this for any distance desired.

Interesting indeed is the method by which the unattended relay stations may be turned on or off from either one of the terminal stations by radio. The receivers at each of the four stations are always alive and ready to catch impulses from their assigned transmitters. When it is desired to make the circuit ready for traffic, New York or Philadelphia starts up its transmitter and sends a certain musical note which the receiving circuits are pre-set to "recognize." At the unattended receiver at New Brunswick, for instance, the tone passes through electric filters somewhat like a key passes through the tumblers of a lock. Electrical circuits "accept" the tone and relays are actuated, turning on the power for the "south" transmitter, which, when in operation, passes the tone on by radio to the

Arney's Mount station. There the operation is repeated. When the tone reaches the Philadelphia station, the transmitter at that city is also automatically turned on, and the tone starts on its return journey back to New York.

(Fig. 83) (Fig. 82)

Fig. 82. Receiver Employed in R.C.A. Ultra-Short Wave Radio Circuit between New York and Philadelphia

Fig. 83. Rear View of R.C.A. Ultra-Short Wave Receiver

Operators in New York know that when the tone comes back to them from the "north" transmitter at New Brunswick the entire circuit is in full operation and ready for traffic. The constant presence of the tone keeps the relay closed, and the circuit is in operating condition. When the tone is withdrawn from the circuit, relays click in the same succession over the round trip to Philadelphia, and, one by one, the transmitters are automatically turned off. Philadelphia has the same control over the circuit as New York.

Antennas, used in this circuit, because of their curious form, have been characterized as "Christmas trees" and "turnstiles." Odd shapes of certain parts of receivers and transmitters result from the application of the principle of "resonant lines" to both transmitters and receivers. This principle, developed by R.C.A. engineers for this use, eliminates crystal control and provides economical and efficient means of maintaining radio equipment in steady tune at extremely short wave-lengths.

The heart of the receiver is the "shoe button" or "acorn" tube, so called because of its small dimensions, and in the transmitters there are new power tubes specially designed for microwave service.

This circuit, as now developed, enables the transmission of drawings, type matter, handwriting, and other visual material in facsimile, along with the simultaneous operation of automatic typewriter and telegraph channels. It is completely secret, in that if the composite signal were "picked up" by any other than the designated receivers, it could not be "unscrambled" into its component parts.

Facsimile transmission does not, of course, fall within the scope of electronic television, for the scanning is mechanical and there are many other divergent features; yet this circuit has proved that the obstacle of distance in ultra-high frequency transmission has been overcome.

Thus, it will again be seen, that electronic television is not a thing apart, since the ultra-high frequency radio circuits that might, in the future, be used for television are also highly useful for other means of communication.

When the ultra-high frequency radio circuit between Philadelphia and New York was opened, David Sarnoff, president of the Radio Corporation of America said, "Radio communication is today placing in useful public service, a region of the radio spectrum which only yesterday was virtually unexplored and scientifically unconquered territory. Having developed a technique of operation for the three meter band of radio wave-lengths, we find in that region, a medium of transmission unlike anything that we have known."

INDEX

INDEX

K

L

M

O

P

R

S

TELECOMMUNICATIONS

An Arno Press Collection

Abramson, Albert. **Electronic Motion Pictures:** A History of the Television Camera. 1955

[Bell, Alexander Graham]. **The Bell Telephone:** The Deposition of Alexander Graham Bell in the Suit Brought By the United States to Annul the Bell Patents. 1908

Bennett, A. R. **The Telephone Systems of the Continent of Europe** and Webb, Herbert Laws, **The Development of the Telephone in Europe.** 1895/1910. Two vols. in one

Blake, George G. **History of Radio Telegraphy and Telephony.** 1928

Bright, Charles. **Submarine Telegraphs:** Their History, Construction and Working. 1898

Brown, J. Willard. **The Signal Corps U. S. A. in the War of the Rebellion.** With an Introduction by Paul J. Scheips. 1896

Chief Signal Officer, U. S. Signal Corps. **Report of the Chief Signal Officer: 1919.** 1920

Danielian, N[oobar] R. **A. T. & T.:** The Story of Industrial Conquest. 1939

Du Moncel, Count [Theodore A. L.] **The Telephone, the Microphone, and the Phonograph.** 1879

Eckhardt, George H. **Electronic Television.** 1936

Eoyang, Thomas T. **An Economic Study of the Radio Industry in the United States of America.** 1936

Everson, George. **The Story of Television:** The Life of Philo T. Farnsworth. 1949

Eyewitness to Early American Telegraphy. 1974

Fahie, J[ohn] J. **A History of Electric Telegraphy to the Year 1837.** 1884

Federal Communications Commission. **Investigation of the Telephone Industry in the United States.** 1939

Federal Communications Commission. **Public Service Responsibility of Broadcast Licensees.** 1946

Federal Trade Commission. **Report of the Federal Trade Commission on the Radio Industry.** 1924

Fessenden, Helen M. **Fessenden:** Builder of Tomorrows. 1940

Hancock, Harry E. **Wireless at Sea:** The First Fifty Years. 1950

Hawks, Ellison. **Pioneers of Wireless.** 1927

Herring, James M. and Gerald C. Gross. **Telecommunications:** Economics and Regulations. 1936

Lodge, Oliver J. **Signalling Through Space Without Wires:** Being a Description of the Work of Hertz and His Successors. [1900]

McNicol, Donald. **Radio's Conquest of Space:** The Experimental Rise in Radio Communication. 1946

Plum, William R[attle]. **The Military Telegraph During the Civil War in the United States.** With an Introduction by Paul J. Scheips. Two vols. 1882

Prime, Samuel Irenaeus. **The Life of Samuel F. B. Morse,** L.L. D., Inventor of the Electro-Magnetic Recording Telegraph. 1875

The Radio Industry: The Story of Its Development. By Leaders of the Radio Industry. 1928

Reid, James D. **The Telegraph in America:** Its Founders, Promoters and Noted Men. 1879

Rhodes, Frederick Leland. **Beginnings of Telephony.** 1929

Smith, Willoughby. **The Rise and Extension of Submarine Telegraphy.** 1891

Special Reports on American Broadcasting: 1932-1947. 1974

Thompson, Silvanus P., **Philipp Reis:** Inventor of the Telephone; A Biographical Sketch. 1883

Tiltman, Ronald F., **Baird of Television:** The Life Story of John Logie Baird. 1933

Wile, Frederic William. **Emile Berliner:** Maker of the Microphone. 1926

Woods, David L., **A History of Tactical Communication Techniques.** 1965